To Save the Phenomena

To Save the Phenomena

AN ESSAY ON THE IDEA OF PHYSICAL THEORY
FROM PLATO TO GALILEO

Pierre Duhem

Translated from the French by
Edmund Dolan and Chaninah Maschler

With an Introductory Essay by Stanley L. Jaki

The University of Chicago Press
Chicago and London

Translated from the French text, originally published as "ΣΩΖΕΙΝ ΤΑ ΦΑΙΝΟΜΕΝΑ: *Essai sur la notion de théorie physique de Platon à Galilée,"* Annales de philosophie chrétienne, *79/156 (ser. 4, VI), 113–38, 277–302, 352–77, 482–514, 576–92. In 1908 it was reprinted under the same title by A. Hermann et Fils.*

The University of Chicago Press, Chicago 60637
The University of Chicago Press, Ltd., London WC1

Midway reprint edition 1985
Printed in the United States of America

94 93 92 91 90 89 88 87 86 85 5 4 3 2 1

ISBN: 0–226–16921–9

Contents

Translator's Note

Brother Edmund Dolan and I independently completed translations of Duhem's *Essai.* He kindly placed his own translation at my disposal, and it substantially improved my version. He is not, however, responsible for errors.

C.M.

Introductory Essay

STANLEY L. JAKI

In 1908 a series of five articles appeared in the *Annales de philosophie chrétienne*, a periodical of rather limited appeal. The titles of the articles also, with their very scholarly air, indicated anything but the expectation of widespread attention. Yet no sooner had the last of the articles been printed than all of them were brought out as a separate volume by the prestigious Parisian publishing house A. Hermann et Fils. Such haste was obviously a recognition of the intellectual stature of the author, Pierre Duhem (1861–1916), professor of theoretical physics at the University of Bordeaux. Clearly, what Duhem wished to convey in those articles about the meaning of physical theory seemed worth bringing to the attention of scholars all over the world.

French was then the leading vehicle of worldwide intellectual communication, and even now with English taking its place as the lingua franca, familiarity with it remains an essential ingredient of scholarship. It is therefore natural to assume that the translation of a fifty-year-old work, and from French to English, aims at reaching more than an academic audience. Scholars can readily avail themselves of the original, as most major libraries list a copy of Duhem's little book with its esoteric main title in Greek, ΣΩΖΕΙΝ ΤΑ ΦΑΙΝΟΜΕΝΑ, and with its scholarly subtitle, *Essai sur la notion de théorie physique de Platon à Galilée.*

Today the time is ripe for broad exposure to a central aspect of Duhem's thought, and the English translation of this book will admirably serve that purpose. The message in its widest perspective is cultural. It was not, however, spelled out by Duhem in one of those popular essays on culture and science which as a rule are more productive of profit than of true enlightenment. Duhem's all-important message is embodied in highly learned publications on the history and philosophy of science which analyze with consummate mastery the early and recent phases of scientific conceptual development—not that Duhem considered himself either a historian of science, though he was undoubtedly one of the greatest, nor a philosopher, though his was a philosophical acumen of rare penetration and finesse.

In his own eyes Duhem was always a physicist, even when engaged in his monumental studies on the history and philosophy of physics. Fame and recognition in these fields came to him more readily than in physics, where he labored with no less distinction. But he was never to receive the highest form of recognition accorded French physicists of his time, a chair of physics in Paris. Ironically enough, he need only have indicated his availability to receive the freshly created professorship of the history of sciences at the Collège de France. Yet he chose to pass over the opportunity with the remark: "I am a theoretical physicist, and either I shall teach theoretical physics in Paris or I will not return there."

He never did. Following the conclusion of his studies and research at the Ecole Normale he lived in "exile," holding teaching positions in various provincial French universities. His third teaching job, the professorship of theoretical physics at the University of Bordeaux, which he obtained following rather brief stays at the science faculties of Lille and Rennes, was also his last. It was from Bordeaux that he tried to influence the course of physics as it emerged from its classical phase. Although he could write books and articles in Bordeaux, the best students were at Paris, and there also was the scientific "establishment," whose consensus could hardly be swayed from the provinces. Nor were most influential scientists in Paris willing to listen to Duhem. Actually Duhem's views were opposed by a wall of silence that had been carefully raised from the moment he submitted his doctoral dissertation. Under normal circumstances a brilliant future would have been in store for the twenty-three-year-old author of a dissertation completed before he had obtained his *licence*. The dissertation dealt with the application of thermodynamic potentials to problems in chemistry and electricity. More important, it also disproved the maximum-work principle, a favorite theorem of the chemist Berthelot, who for the previous twenty years had been trying to develop it into a cornerstone of physical chemistry. Duhem's thesis was taken by Berthelot as a personal offense, and it was owing to his influence that it was rejected.

Serious as the blow was, it can hardly be considered fatal, for within two years Duhem's talents produced another dissertation, a mathematical analysis of the theory of magnetism. More damaging to Duhem was a disconcerting aspect of scientific life which sometimes prevents the triumph of scientific conclusions however well argued. Scientists like others have their foibles and sensitivities and can resort on occasion to unsavory tactics to protect their own renown and position, even at the expense of truth and progress. Berthelot's attitude toward Duhem was a case in point. Moreover, his was a pivotal position in the power structure of the strongly centralized French scientific establishment. He served as inspector general of French higher education, as minister of public education, as minister of foreign affairs, and from 1889 he held the prestigious post of permanent secretary of the French Academy. No wonder that enduring obeisance was paid to his revengeful verdict on Duhem, the doctoral candidate: "This young man shall never teach in Paris." These words also crystallized the resentment of several prominent personalities who were patently jealous and suspicious of Duhem's talents, energy, and uncompromising character. Thus even after Berthelot had come to relent, Duhem, in the executive circles of French higher education, was still branded as a dangerous nonconformist to be kept at arm's length. Or as Jules Tannery, a friend of Duhem, once summed up the situation to him: "Being disagreeable to big shots had been made part of your identity."

Rightly or wrongly Duhem was a thorn in the side of many. The best aspects of his character were also the ones that made him enemies. His brilliance, combined with utter honesty, selfless dedication, and crusading verve, earned him not only the unreserved admiration of his students (they found in him "a teacher who cared") but also the resentment of many of his peers. Needless to say, his strongly conservative political views and his deep Catholic convictions could bring him no favor in the liberal and anticlerical atmosphere of the Third Republic. His readiness to uphold right causes set him on more than one occasion at loggerheads even with friends. An especially sacred cause in his eyes was the purity of scientific truth, which he saw threatened by the fallacies and contradictions of mechanism. In fact he rated the search for a mechanical explanation of the universe "the most dangerous stumbling block for theoretical physics." However much truth the statement contained, its sweep could only alienate most of those who like Jean-Baptiste Perrin considered the early triumphs of atomic physics a vindication of Descartes' mechanistic interpretation of nature.

To be fully aware in the 1880s of the breadth and width of the mechanistic fallacy in physics required unusual talents and independence of opinion. After all, by then the identification of the intelligible with the mechanical had been a fundamental article of the scientific creed for more than two hundred years. True, before Duhem, Lagrange powerfully steered the science of mechanics away from the shallows of mechanistic imagery. Ampère's work in electricity also showed to good advantage the purely formalistic aspects of mathematical physics. As early as 1855, Rankine, a pioneer in thermodynamics, spoke of a science of energetics, designed to achieve a thorough demechanization of physical theory. By the time Duhem received his doctorate in 1888, Mach had been pursuing for almost two decades his unrelenting critical analysis of the conceptual development of classical mechanics.

All of these were, however, partial efforts. Only Duhem had the courage, stamina, and talent to undertake on a broad front a radical recasting of theoretical physics. The true measure of his efforts can best be seen in that impressive *Notice* which he submitted to the French Academy prior to his election as one of its first six nonresident members. This was in 1913, only three years before his death at the untimely age of fifty-five. The *Notice* of 130 printed pages consists of two parts. The first is a list of his publications running over thirty pages, striking evidence of his gigantic output. If the publications of his last three years are added, the total constitutes some thirty books and nearly four hundred articles. For anyone interested in a fully authentic account of Duhem's thought, the second part of the *Notice* is a priceless gem. There, in over a hundred pages, Duhem offers an analysis and summary of his aims, motivations, and accomplishments in theoretical physics and in studies related to the philosophy and history of science.

Only twenty pages of the *Notice* are devoted to his philosophical and historical writings—a genuine reflection of Duhem's self-evaluation. The first eighty pages of the document deal with his research in theoretical physics. These he lists under the following headings: (1) codification of the principles of energetics, (2) mechanics of fluids, (3) mechanics of elastic bodies, (4) chemical mechanics, (5) equilibrium and motion of mixed fluids, (6) friction and false equilibria, (7) permanent changes and hysteresis, (8) galvanism, magnets, and dielectric bodies, (9) electrodynamics and electromagnetism.

The scientist most interested in tangible results that "work" in the laboratory and industry will appreciate Duhem's pioneering analysis of shock waves and their modifications in viscous fluids. Duhem's *Recherches sur l'hydrodynamique* took on a special importance with the coming of supersonic flight and was reprinted in 1961 by the French Ministry of Air. Theoreticians of the liquid state will value his original ideas on systems equivalent to liquid crystals. Thermodynamicists will set store by his axiomatization of the principles of thermodynamics. Today it is generally recognized that Duhem ranks with Ostwald, Arrhenius, Van't Hoff, and Le Chatelier as a founder of physical chemistry. Actually, Duhem's scientific work began with a sustained effort to recast the theoretical foundations of chemical processes on the basis of a generalized thermodynamics. His main inspiration was the American physicist, Joshua Willard Gibbs of Yale, and it was in part through Duhem's work that Gibbs gained recognition in Europe.

Duhem intended to include all major branches of physics in his generalized thermodynamics. In the case of mechanics, hydrodynamics, and elasticity he had no difficulty. There the analogy between the potential of classical mechanics and Gibbs's generalized thermodynamic potentials is close enough. Electrodynamics is another matter. There Duhem found other stumbling blocks as well. Maxwell's electromagnetic theory, which increasingly dominated the scene, had some highly irritating aspects of its own for the resolutely antimechanistic Duhem. Even in its final form, with the scaffolding of mechanical models removed, Maxwell's theory still betrayed its mechanistic moorings. In view of Duhem's crusading character it was almost inevitable that he should undertake a comprehensive critique of the work of the great Scotsman. This he did in his *Théories électriques de J. Clerk Maxwell: Etude historique et critique*, published in 1902. But much of what there was to criticize in Maxwell's great papers consisted of trivia—mistaken signs, algebraic errors, inconsistencies in using terms, and the like. By insisting on such details Duhem merely showed the Frenchman's passionate love of a sweeping, unitary logic, which he liked to extol over the pragmatic Anglo-Saxon approach. According to Duhem, Maxwell's work was a classic embodiment of the latter. Following his compatriot Poincaré, Duhem described Maxwell's theory as lacking the characteristics of a single, definite, well-organized edifice. In his eyes, it was rather a group of provisional buildings with passageways between them that were tenuous or sometimes nonexistent. It was Duhem's firm belief that his own three-volume *Leçons sur l'électricité et magnétisme*, (1891–92) presented a more logical alternative. To some extent this was true. Duhem's work, which relied heavily on contributions by Helmholtz and Carl Neumann, did better justice to some of Poison's fundamental ideas than did Maxwell's discussion, which rested on another tradition, namely, the one leading from Boscovich to Faraday. But on a most crucial point, the inclusion of electromagnetic waves, Duhem's procedure had to buy success at the price of obvious inconsistencies and cumbersome complications.

Duhem was fully aware that his was a lonely voice. In the whole year of 1913, as he remarked in a letter to his daughter, only one copy of his massive opus on electricity was sold. Not that he admitted defeat. He felt certain that the future course of physics would run much along the lines outlined in his *Traité d'énergétique* (1911), which he considered his most significant contribution to physics. A baffling appraisal indeed, for throughout the two heavy volumes of that work there was not a single reference to atoms. By 1911 radioactivity was more than a dozen years old, and Ostwald, a leader of the energeticist school, was just beginning to admit the existence of atoms. A few years later, with Duhem still alive, Mach's resistance to atoms also caved in—and under dramatic circumstances. "Now I believe in the existence of atoms," he said to Stefan Meyer, a Viennese physicist, when he brought to the ailing Mach's bedside a scintillation screen and let him marvel at the little stars of light produced on it by a speck of radium. Duhem held out to the very end. He saw no need to modify the fundamental principle of his methodology, that physical theory should not contain speculations about the properties of a layer of matter underlying the phenomena. By that layer Duhem meant of course the ultimate layer, which at that time atoms were believed to represent.

During the heady days of the first phase in the discovery of the fundamental particles, Duhem's intransigence could easily be classified as an old man's prejudice about the true nature of the existing situation. Yet he was not so wrong as he appared to be in the eyes of other physicists of the day. Not only did the atom, the "indivisible," turn out to be composed of parts, it soon became evident that its massive center too was glued together of smaller components, protons and neutrons. The proton itself as the "primordial particle" has since given more than one hint of its complex nature. The presumed fundamental trinity constituted by the proton, neutron, and electron has also had to yield as science caught, with the discovery of the positron, its first glimpse of the realm of antimatter. During the last thirty years physics has seen a bewildering multiplication of "fundamental" particles that are such in name only. Indeed, there is a growing feeling that some of the new particles may be not so much nature's as man's products. At any rate, the fundamental particles of modern physics are not particles in the sense that atoms were believed to be particles in Duhem's time. And when physics tries to account for the interaction of its charged fundamental particles with the electromagnetic field, it no longer relies on the standard procedure of postulating an underlying medium of interaction. The seat of these interactions, the empty space of quantum electrodynamics, is a mathematical fiction whose sole function is to "save the phenomena."

This is not to suggest that Duhem's methodology has therefore been simply vindicated. That methodology bears in more than one point the irremediable shortcomings of its times and of its author. It would, indeed, be folly to follow Duhem's often rigid precepts in methodology. Applying them in full vigor would mean the renunciation of the search for more particles and the scrapping of plans for accelerators much larger than the existing ones. Although the succeeding generations of accelerators have failed to bring man within reach of the fundamental layer of matter, they have given him convincing proof of nature's awesome richness and complexities. Yet the results also indicate that the model-making, particle-hunting sector of physical research is not likely to become a fully successful, self-consistent enterprise. It seems decisively incapacitated by that strain of naïve realism which Duhem as teacher, theoretician, philosopher, and historian of science tried to banish altogether from physics.

Duhem's crusading opposition to all manifestations of naïve realism in physics soon earned him the positivist label. Today he is all too often lumped together with Comte, Mach, the operationalists, and even with the logical offshoot of these latter, the fallibilists. Duhem's positivism is, however, a very qualified one. It certainly has nothing fundamental in common with that aspect of Comte's positivism which turns the latter into a pseudometaphysics, if not a pseudotheology. Against Mach, the operationalists, and the fallibilists, Duhem firmly believed that the human mind has the ability to learn something about the true, inner nature of the physical world, though not by relying exclusively on the quantitative method. What Duhem clearly perceived—and this was no small accomplishment, considering the scientific philosophy of his time—was that the quantitative, formalistic approach did not suffice even in physical science, let alone in other areas of human inquiry. Contrary to the typically positivist precepts, physicists have always done their experimental research in the belief that their work related to physical reality and that their conclusions and laws were not merely convenient formulas to be reshuffled at will, but rather that they revealed something, however little, of nature.

Duhem would find it gratifying that today the word *faith* is frequently on the lips of leading physicists when they try to account for the ultimate source of their confidence in their work. By faith is not meant, of course, any attachment to supernatural propositions but rather an acknowledgment of the indispensability of man's intuitive powers. It is through these powers and not through simply discursive quantitative reasoning that man gets hold of aspects of reality hidden behind the realm of the phenomena. Among these aspects are the unity, simplicity, symmetry, and uniformity of nature, and without a firm belief in them the most vital fiber of the scientific enterprise would become atrophied. As a justification Duhem could have recalled countless details of scientific history, but he preferred to rest his case on an existential affirmation that he considered to be beyond any formal proof or disproof. In this connection he liked to recall the words of his favorite philosopher, Pascal, whose *Pensées* he knew almost by heart: "We are impotent in proving, and this impotence cannot be conquered by any dogmatism; we have an idea of truth which cannot be conquered by any Pyrrhonian skepticism."

A corollary to Duhem's views on physical reality, physics, and intuition was his contention that as physics progresses it approaches gradually, though asymptotically, that ultimate and solely valid form of physical theory which he called the "natural classification of phenomena." In support of this conviction, which once more sets him radically apart from most operationalists and certainly from all fallibilists, Duhem reached back characteristically enough to existential grounds, to the realm of the irreversible, unique *events* of scientific history. Their sequence presented, Duhem believed, a broad display of both fruitful and misleading approaches. Reflecting on the evidence, the physicist was therefore in a position to recognize the proper guidelines, which physical theory as such could not provide. "Physics is not capable of proving its postulates, nor does it have to prove them" was Duhem's own precept, which he stated bluntly in his *Notice* to the Academy. What could not be done by the philosophy of physics could be achieved by a judicious reading of its history.

As a reader of the history of physics Duhem had few equals. To vindicate his energetics, as he called his own brand of physical theory, Duhem produced two major works, *L'évolution de la mécanique* (1903) and *Les origines de la statique* (1905–6). Studying the evolution of mechanics implied of necessity a close look at its origins or at the science of Galileo and his times. Around 1900 most students of the question would have simply opened up the discussion with Galileo, but Duhem, with the true instinct of a born historian, knew that in intellectual history the beginnings are rarely abrupt. Pursuing his hunches Duhem was led beyond Leonardo's science to a previously untapped and largely forgotten collection of medieval manuscripts gathering dust in the Bibliothèque Nationale in Paris. What he found there revolutionized the history of science. Singlehanded he destroyed the legend of the "scientific night of the Middle Ages." Before him, the phrase was a hallowed shibboleth of a self-styled Enlightenment. After him it has become the sign of an inexcusable ignorance which unfortunately lingers on. Thus a recent and widely read *Biography of Physics* from the pen of a noted physicist completely ignores medieval science with the remark that "primitive Lysenkoism" was then flourishing all over Europe.

Duhem's historical investigations on the origins of statics opened up for him the fascinating world of ancient Greek science, and with it the first phase in that great continuity which Duhem saw in the evolution of physical science. From that point on Duhem had an added motivation in his historical researches. He still had a keen eye for any past evidence of that formalistic approach which constituted the backbone of the methodology of his energetics. But he also became aware of his mission to set straight the record of the history of physical science. The new record as worked out by him was nothing short of monumental. Between 1906 and 1913, he gathered in three volumes his studies on Leonardo, on Leonardo's sources, and on those who during the sixteenth century learned their physics, the physics of medievals, from Leonardo and his sources. (*Études sur Léonard de Vinci: Ceux qu'il a lus et ceux qui l'ont lu*). As most sixteenth-century scientists were eager to follow their humanist forebears in decrying the alleged backwardness of the thirteenth and fourteenth centuries, they kept silent about the true fountainhead of their information. It requires the painstaking investigation of the historian to show that they often quoted verbatim from the writings of medieval men of science.

While Duhem's studies on Leonardo had already given a strong indication of the wealth of the medieval material, its true richness was first revealed in a systematic manner in the monumental *Système du monde*, the publication of which started in 1913. In that work Duhem wanted to give a detailed account of the development of physical theory from the pre-Socratics to the birth of classical phyics. He surely felt that he was writing the crowning work of almost two decades of trailblazing historical research. The rate at which he must have worked is truly astonishing. In the span of four years he completed, without collaborators, the manuscript of ten of the planned twelve huge volumes, and he also succeeded in sending the first five volumes into print. He clearly looked beyond the hour of completing the enormous undertaking. "When I have finished my *Système du monde*," he kept telling friends, "I will seclude myself during the vacations at Cabrespine, and I will spell out its essential conclusions in a work of three hundred pages free of scholarly apparatus." It was not given to him to carry out this cherished plan, although it was uppermost in his mind even during the last two weeks of his life, when his strength was rapidly ebbing away after a sudden heart seizure in early September 1916.

Duhem's insights into the history of physics were, however, sufficiently embodied in the first five volumes of the *Système du monde* to assure its lasting impact. As Duhem rightly emphasized, before the seventeenth century only one part of physics, astronomy, had achieved that degree of development where mathematical theory and experimental observation were in a meaningful interaction. This is why in the *Système du monde* a prominent place is given to ancient Greek and medieval astronomical and cosmological theories. Fully aware of the richness of the medieval material, Duhem devoted much of his lengthy work to a discussion of the scientific writings of medieval scholars. Actually, about one-third of volume two forms the opening of the second major part of the work, the discussion of medieval astronomy. The third part, the rise of medieval Aristotelianism, is in volumes four and five. The next four volumes, containing the fourth and fifth parts of the work, are devoted to the decline of Aristotelianism and the emergence of the rudiments of a new physics at the University of Paris during the fourteenth century. Since Duhem, in a number of articles published from 1895 on, had aroused interest in that last and most important topic, the medieval origins of classical physics, it was only natural that the manuscript of these posthumous volumes should sooner or later go into print. By the 1950s research in medieval science had become a thriving field that seemed to justify the efforts and risks of such a publishing venture. The growth of medieval research not only kept alive interest in Duhem's ideas but also produced worthy successors such as Anneliese Maier, Ernest Moody, Alistair Crombie, Marshall Clagett, and others. Their investigations considerably modified some of Duhem's conclusions—conclusions which he himself stated more moderately in the *Système du monde* than in his *Etudes sur Léonard de Vinci.*

It is now generally agreed that Duhem gave too much weight to the condemnation of some Averroist theses by Archbishop Tempier of Paris in 1277. In the solemn striking down of theses that limited the omnipotence of God to the creation of a typically Aristotelian cosmos, Duhem saw a decisive prompting of medieval scholars to speculate freely on the possible configurations and laws of the physical world. Duhem's overemphasis of the scientific significance of a theological decision was in part due to his religious sympathies. Yet it is well to remember that these sympathies greatly helped him in sighting a vast field which many scholars before him, and some even after him, had systematically ignored because of their very different sympathies. Again, while Duhem was right in emphasizing the indebtedness of Galileo and his successors to late medieval science, it remains true that their refinements of the impetus theory still should be considered crucial. Some particularly favorite theses of Duhem about Oresme also proved untenable. Oresme clearly cannot be considered an inventor of analytical geometry and an early proponent of the rotation of the earth. Subsequent historical research also has supported neither Duhem's emphasis upon the role of Jordanus Nemorarius nor his speculation about a presumed precursor of Leonardo. On careful reflection one also finds that Duhem painted an overenthusiastic picture of the intellectual vigor of the University of Paris in the age of Buridan. Here, Duhem's patriotism was clearly in play.

Patriotism is also evident in his failure to recognize the importance of Thomas Bradwardine's reformulation of the so-called Peripatetic law of motion and in his unappreciative treatment of the contributions of Oxford's Merton College. This patriotism of Duhem was a distinct factor, it may be noted, in his opposition to nineteenth-century British physics bent on model-making. It also inspired two small wartime books. In one of them, *La chimie est-elle une science française?*, he proudly took issue with Ostwald's claim that chemistry was a German science. In the other, *La science allemande*, he made the most of the survival in German thinking of some obscurantist and debilitating traces of Naturphilosophie. Yet for all their propagandistic flavor, both these works are remarkably free of the vicious invective that fills the pages of that wartime literature which so many French and German scientists embellished with "scientific" details.

Duhem's religious sympathies and patriotic feelings can be found at work even in *To Save the Phenomena* without, however, invalidating its main conclusion. In a sense this relatively short book may be considered an authentic capsule version of the huge volumes of the *Système du monde.* The richness of detail in the latter can indeed be seriously distracting to the reader unaware of Duhem's main objective in analyzing the history of ancient and medieval astronomy and physics. In Duhem's huge opus it is by no means easy to keep track of the theme which he considered the principal lesson of studies in scientific history: the recognition of the leading role of formalistic constructs in science as against the realistic interpretation of theories. The reader of *To Save the Phenomena* can hardly complain that its author has not been explicit enough about the main message of the book. From the outset he is reminded again and again that Plato's definition of the aim of astronomy, "to save the phenomena," has been the most seminal, the most sensible, and the most logical guiding principle in scientific speculations as centuries of scientific history have gone by.

The work is largely a documentation and interpretation. As a documentation it contains the most relevant texts from Plato's to Galileo's times about the possible formulations of physical (astronomical) theory. Although these texts can be found in various scholary pubications, the present English translation provides an easy access to them not available elsewhere. The texts illustrate two main traditions in astronomical investigations, the formalistic and the realistic approaches. The former, originating with Plato, considered the various geometrical models of planetary motions and of the construction of the cosmos as mathematical expedients. The homocentric circles, the epicycles, the deferents, the extants do not correspond in this view to the wheels of an intricate mechanism. Rather they are geometrical patterns that facilitate calculations necessary for finding the position of planets at any given moment. On the other hand, the realistic interpretation of astronomical theory assigned physical reality to these geometrical patterns. Consistency then demanded that only those aspects of the patterns be retained which did not conflict with the physical, which also meant common-sense reasoning.

In classical antiquity the realistic school was represented by Aristotle, Posidonius, Theon of Smyrna and Simplicius, to mention only the principal names. According to their position the basic tenets of physics dictated the choice between the various geometrical or mathematical representation of celestial motions. The most important of these basic tenets was the "naturalness" of circular motion. Clearly this was not so much a postulate of physics, as we understand the word *physics* today, but rather a postulate which derived from considerations relating to a realm *beyond physics,* that is, *metaphysics.* In other words, in the realistic interpretation the truth of astronomical (physical) theory depended on the truth of that particular philosophy which provided the rules of selection among various explanatory devices.

In the formalistic approach astronomical or physical theory was not subject to metaphysical considerations about the physical. It could adopt any geometrical or mathematical procedure, since its sole purpose consisted in calculating (or saving) the occurrence of a given celestial phenomenon, be it an eclipse, an opposition, an aphelion, a retrogression, the rate of advance of a planet along the belt of the zodiac, or whatever. The chief representative of this approach in classical antiquity was Ptolemy. His systematic use of the eccentric was not only diametrically opposed to the system of homocentric, crystallike spheres—that purest form of the realistic interpretation of the system of planets—it also flew in the face of any attempt to construct a workable mechanical model for it. Such difficulties caused Ptolemy little if any concern. He boldly claimed that astronomical theory was to have only two qualifications: it should yield good numerical results (thereby saving the phenomena), and its geometrical apparatus should conform to the rule of greatest possible simplicity.

The rule of greatest simplicity was warmly espoused by the great Jewish scholar Maimonides, the only important figure among medieval Arabic and Jewish philosophers to side with Ptolemy's formalistic approach in asronomy. The popularity of Aristotle among the Arabs inevitably produced a realistic interpretation of astronomical theories, with Averroes and al-Bitrogi in the vanguard. Christian medieval students of astronomical (physical) theory followed for the most part a line of compromise. While acknowledging the truth of Aristotle's physics, they also admired the precision of the mathematical procedure, contrary though some of its assumptions were to Aristotle's physics. Theirs was obviously a heterogeneous set of attitudes characterized by an inability to make a decisive case either for the formalistic or for the realistic approach in astronomy. Duhem therefore overstates the case, at the same time revealing his strongly promedieval sympathies, when he writes that the scientific philosophy of Christian astronomers in the Middle Ages reduced to two principles: greatest simplicity and greatest exactitude. The actual situation, as the texts quoted by Duhem show, reveals a great deal of uncertainty and hesitation, as well as a lack of thorough familiarity by medieval scholars with Greek astronomical theory.

The disputation of the respective merits of the formalistic and realistic methods in astronomy took on a far greater vigor and depth as the Renaissance reached its zenith around 1500. It was then that young Copernicus studied in Italy and witnessed at close range the passionate quarrel between two groups of Italian astronomers: one, the Averroists, who despised the mathematical approach, the other, astronomers with Pythagorean preferences, who attributed to it a realistic (we would perhaps say today, a heuristic) value. Obviously, both were extremist positions, and this gave Duhem an opportunity to extol the "well-balanced" position of the leaders of the Parisian school. In fact he credits them with the insight really spelled out by Nicholas of Cusa, that both the superlunary and sublunary regions obey the same laws (of physics). From the beginning of the fourteenth to the early fifteenth century, Duhem claimed, the University of Paris had been voicing propositions about the method of physics, the correctness and depth of which far surpassed all that the world was to hear in that regard until the mid-nineteenth century.

For all the exaggeration and lopsidedness of such a claim, its grain of truth serves as a good background against which the Copernican problem should be viewed. Around the turn of the century, when Copernicus was still generally regarded as a champion of experimental method, it took scholarship and independence of mind to point out that Copernicus was a mathematical realist. In other words, Copernicus saw in the geometrical simplicity of the heliocentric arrangement of planets a convincing proof that such indeed ought to be be the case. This appraisal of the demonstrative strength of mathematical (geometrical) analysis of the phenomena was heartily echoed by Rheticus and became increasingly the hallmark of Copernicans, as amply illustrated in Kepler's and Galileo's writings. Kepler's mystical belief in the power of numbers and Galileo's unbounded admiration for Pythagoras bring out clearly the basic feature of the "new physics" in which the systematizing and predictive value of the mathematical apparatus served as an irrefragable proof of its one-to-one correspondence to the physical reality. A most telling description of this was given by Galileo when he repeatedly praised Copernicus in the *Dialogue* for having the courage to set (mathematical) reasoning over the evidence of his senses. As the mathematics used in physics at that time was largely couched in geometrical patterns, it was only natural to assume that the circles, triangles, and squares reflected pieces of mechanism. Thus was born the naïve realism of mechanistic or classical physics which constituted the dominating scientific atmosphere for centuries. Only the acumen of a few critical minds could break through the fog of fallacy in which intelligibility was equated with machinery. Duhem was one of the few.

He certainly succeeded in diagnosing an all-important issue that separated astronomers into two camps during the period that started with the publication of Copernicus' great work and reached its climax in Galileo's condemnation. Locked in conflict were two misguided realisms: the mathematical realism of the Copernicans and the naïve realism of Peripatetic philosophers who, to make matters worse, also cited passages of the Bible as criteria of the physical truth. The main representatives of the latter trend among astronomers were Tycho and Clavius. The chief spokesmen of the former were of course Kepler and Galileo. Some, like Bellarmine, tried to prevent the conflict from coming to a head. Significantly, Bellarmine reached back to the view of Ptolemy paraphrased in Osiander's preface to the *Revolutions*. That this third position appeared to Duhem's positivism the most judicious stance that could be taken in the matter is only natural. Actually the situation resembled a triangle, each of its corners being more or less equally removed from the truth at the center. Not that in the context of the times this could be spelled out in a convincing manner. Each camp had its share of strong and weak points, and all lacked what only three centuries of scientific development could provide, perspective. The Copernicans lacked the science of dynamics needed to cope with the enormous physical problems created by the moving earth. The common-sense philosophers could not talk away the overwhelming attractiveness and effectiveness of mathematics (geometry) in dealing with the phenomena. Advocates of the aloofness of Ptolemy's method could not be convincing in the face of men's invincible faith that it was possible to come to grips with physical reality. It took time for dynamics to mature. Even more time was needed to let mathematics reveal some of its inherent limitations. Again, only time showed that for all the spectacular probing of physics into the heart of matter, man's ties with nature hinged ultimately on an act of philosophical faith.

All this can clearly be seen today, but this takes a careful reading of the development of physical theory, including its all-important twentieth-century phase, of which Duhem at best caught only an early glimpse. Naturally, one does not need the vantage point of the 1960s to conclude that the Peripatetics and their allies among astronomers were wrong in their unqualified acceptance of common-sense evidence. Yet for all the recent development of successful physical theories that fly in the face of common sense, the realm of common-sense observation still remains the background against which the truth of all propositions is ultimately judged. Scientific criticism can modify a large number of conclusions based on common sense, but it cannot dispense entirely with common-sense evidence.

As to the mathematical realism of the Copernicans, much of it was vindicated by subsequent developments in science. In classical physics, where model-making determined for the most part the type of mathematical analysis to be used, mathematical realism revealed on several occasions a distinctly heuristic value. Thus the first hint of the existence of conical refraction came from Hamilton's thorough mathematical analysis of double refraction. Again, it was mathematics that revealed to Maxwell that the viscosity of a given gas at a fixed temperature was independent of the pressure. With the advent of quantum mechanics and relativity, mathematics has almost become an "open sesame" to unsuspected areas of physical reality. Mathematical theories developed a hundred years ago, with no eye to physical problems, turned out to be the very formalism needed by relativity and quantum theory. Moreover, the dominating role of selection rules and "magic" numbers in atomic, nuclear, and particle physics lends strong support to the contention that the world is indeed a construct in numbers. Yet at the same time mathematics does not seem to have the ability to formulate its fundamental formalism, which if mathematical realism were unreservedly true, could provide the definitive form of physical theory as well.

Difficulties also beset the third of the contending parties, positivism, whereby mathematical analysis of the phenomena does not bear on the nature of things but is merely a convenient and economical grouping of them. Such an approach in physical science had a liberating effect by helping to throw overboard the unnecessary ballast of many questions and concepts of a metaphysical nature not really needed by physics. At the same time positivism proved to be a lame guide in physics. It certainly did not serve as an inspiration in the exciting search for the atom, the nucleus, and the supposedly fundamental set of particles, which regardless of one's philosophy of science should be considered realities and not mere mathematical formulas. As a matter of fact they are considered realities by the working physicists, most of whom keep a carefully guarded realistic compartment in their thinking in spite of their often vocal advocacy of such modern forms of postivism as operationalism and fallibilism.

What all this suggests is that the search for truth cannot rely on any specific method, be it purely philosophical, religious, or scientific. All these have their limitations, and only a vigorous interplay among them will lead mankind forward on the path of understanding. When any of these approaches takes on an exclusive priority, truth will suffer. Thus when common-sense realism dominated at the expense of quantitative method, as was the case in Aristotelian physics, the investigation of the physical universe became a barren enterprise. When the naïve realism of model-making became invested with the aura of infallibility, belief in the realm of value judgments came to be gradually undermined. And the present-day encroachment of the quantitative method on almost every field of human experience and reflection presents a threat the magnitude of which cannot be overestimated.

Modern culture seems to be in the throes of an unbridled quantification, in which individuals are on the road to becoming mere numbers, if not mere holes in punch cards. As in any crisis, the extremist remedies are here very much in evidence. Side by side with those who decry science as a perversion of "naturalness" are those who want everybody and everything to be ruled by science. To strike a middle course, as sanity demands, between the extremes of romantic primitiveness (if not illusory anarchism) and of dehumanizing scientism, one must be fully aware of the limitations of scientific method. This is not an easy task. To cope with it there are several avenues, of which one, that of historical studies, should have special appeal. History is a great equalizer. Sooner or later it cuts all things and all men down to their true size. Science looms up as a savior only for those whose familiarity with it is restricted to what Duhem so aptly called "the gossip of the moment." Those who are brave enough to look past the popular but ephemeral truths of the day will find in history a most instructive teacher. The history of physical science can indeed forcefully show its student that myths are present in science no less than in other areas that owe so much to science for the reduction of their myths.

Recognition of this may be a humbling experience in a scientific age such as ours; yet it is indispensable if science is to become man's servant rather than his tyrant. Those who pondered much on the proper range of scientific theory and enriched their analysis of it with a wealth of historical illustration have rendered a most valuable service to the cause of culture. Indeed, if the liberating message about the limitations of scientific method is gaining a firm foothold today, a large share of the credit should go to Duhem. His philosophical analysis of the aim and structure of physical theory and (especially) his pioneering studies in the history of science display an increasing timeliness, or rather an enduring humanistic freshness. No wonder. Duhem for all his devotion to scholarly and scientific investigations was visibly animated by a dedication to his fellow men, whom he wanted to assist in their groping toward a more robust, more balanced, and more satisfying formulation of truth. Through the pages of *To Save the Phenomena*, one cannot help sensing the presence of an utterly honest and dedicated mind whose ultimate motivation was far more than gathering academic laurels. His instructive analysis of the age-old scientific program of "saving the phenomena" may therefore be considered a highly relevant cultural contribution. In a genuine sense it is an effort to save from some alluring pitfalls the greatest and most marvelous of all phenomena: the mind of man.

To Save the Phenomena

Introduction

What is the value of physical theory? What are its relations to metaphysical explanation? These are lively questions today, but like so many central questions, they are by no means new. They belong to all time: they have been raised as long as a science of nature has been in existence. The form in which they are cloaked may change somewhat from one century to another; the form of the questions derives from the science of the day and is variable; but one need only remove this covering to become aware that essentially the questions remain the same.

Until we reach the seventeenth century, we come upon very few areas of natural science that have advanced to the point of formulating theories in mathematical language, theories whose predictions are expressed in numerical terms so that they can be verified by comparison with the measurements furnished by precise, direct observation. Even statics, then called *scientia de ponderibus*, and "catoptrics," at that time subsumed under "perspective" (our "optics"), had barely reached this stage of development. Bypassing these two limited areas, we encounter only one science with a form which, even at that time quite advanced, would cause us to anticipate the course taken by our modern theories of mathematical physics: that science is astronomy. Hence, where we today speak of "physical theory," the Greek or Arabic philosophers and the medieval or Renaissance scientists spoke rather of "astronomy."

No other area of natural science had yet reached that state of perfection where the language of mathematics serves to express laws discovered by exact observation. Physics in our sense, as both mathematical and empirical, had not yet become separated from the metaphysical study of the material world, that is, from cosmology. In many instances, therefore, where we would today speak of "metaphysics," the ancients used the word "physics" instead.

This is why the question so much discussed today—What are the relations between physical theory and metaphysics?—was for the two thousand years formulated differently—What are the relations between astronomy and physics?

In the following essay we want to review rapidly the answers given to this question by Greek thought, Arabic science, medieval Christian scholasticism, and, finally, by the astronomers of the Renaissance.

Others, headed toward the same goal as we, have blazed the trail. We could not possibly fail to mention in particular T. H. Martin,[1] Giovanni Schiaparelli,[2] and Paul Mansion.[3] To the texts which they earlier brought to attention we shall be adding a good many others. Together these will, we believe, enable us to reconstitute with some accuracy the conception of physical theory held by philosophers and scientists from Plato to Galileo.

1. T. H. Martin, *Mémoires sur l'histoire des hypothèses astronomiques chez les Grecs et chez les Romains*, pt. 1, "Hypothèses astronomiques des Grecs avant l'époque Alexandrine," chap. 5, par. 4 (*Mémoires de l'Académie des Inscriptions et Belles lettres*, vol. 30, pt. 2).

2. Giovanni Schiaparelli, *Origine del Sistema planetario eliocentrico presso i Greci*, chap. 6 and Appendix (*Memorie del Instituto Lombardo di Scienze e Lettere; Classe di Scienze matematiche i naturali*, vol. 18 [3d ser., vol 9], 17 March 1898).

3. Paul Mansion, "Note sur le caractère géometrique de l'ancienne astronomie," *Abhandlungen zur Geschichte der Mathematik*, vol. 9 (1899).

1
Greek Science

To find the origin of the tradition whose course we mean to follow we must go back to Plato.

The transmission and application of his opinions concerning astronomical hypotheses Plato owes in the first instance to Eudoxus. Next, Eudemus, an immediate disciple of Aristotle, drawing on Eudoxus' writings, reported Plato's views in the second book of his *History of Astronomy*. It was from this book that Sosigenes, the philosopher and astronomer who later became Alexander of Aphrodisias' teacher, borrowed them and passed them on to Simplicius. And it is from Simplicius that we have our report.[4]

In Simplicius' *Commentary* we find the Platonic tradition formulated in the following terms:

> Plato lays down the principle that the heavenly bodies' motion is circular, uniform, and constantly regular.[5] Thereupon he sets the mathematicians the following problem: What circular motions, uniform and perfectly regular, are to be admitted as hypotheses so that it might be possible to save the appearances presented by the planets? (*τίνων ὑποτέθεντων δι' ὁμαλῶν καὶ ἐγκυκλιῶν καὶ τεταγμένων κινὴσεων δυνήσεται διασωθῆναι τὰ περὶ τοὺς πλανωμένους φαινόμενα;*)

The object of astronomy is here defined with utmost clarity: astronomy is the science that so combines circular and uniform motions as to yield a

4. Simplicius *In Aristotelis quatuor libros de Coelo commentaria* 2. 43, 46 (Karsten ed., p. 219, col. a and p. 221, col. a; Heiberg ed., pp. 488, 493).

5. That is to say, always in the same direction.

resultant motion like that of the stars. When its geometric constructions have assigned each planet a path which conforms to its visible path, astronomy has attained its goal, because *its hypotheses have then saved the appearances.*

This is the problem that challenged the efforts of Eudoxus and Calippus: it was to save the appearances (*σώζειν τὰ φαινόμενα*) that they combined their hypotheses. When Calippus modified the combination of homocentric spheres proposed by Eudoxus in certain particulars, he did so solely because the *hypotheses* of his predecessor did not accord with certain *phenomena*, and he was determined that these *phenomena* too should be saved.

The astronomer must declare himself fully satisfied when the hypotheses he has combined succeed in saving the apparances. But may human reason not fairly ask for more? Does it not have the power to discover and analyze some of the characteristics of the nature of the heavenly bodies? And might not these characteristics help him by pointing out certain types to which astronomical hypotheses should of necessity conform? And should not a combination of movements that cannot conform to any of these types therefore be declared unacceptable, though this very some combination would save the appearances?[6]

Along with the *method of the astronomer*, so clearly defined by Plato, Aristotle admits the existence and legitimacy of another such method: he calls it the *method of the physicist.*

In the *Physics*,[7] Aristotle compares the methods of the mathematician and the physicist and lays down certain principles which have direct bearing on the question we just raised, though his remarks do not allow us to push analysis very far. Geometers and physicists, he says, frequently study the same object, whether it be the same figure or the same movement, but they regard the object from different points of view. A particular figure, a movement—the geometer views these "by themselves," abstractly; the physicist, by contrast, studies them as the limit of such and such a body, the movement of such and such a moving thing.

6. The translation is more than free. The French reads as follows: ". . . l'esprit humain n'est-il pas en droit d'exiger autre chose? Ne peut-il découvrir et analyser quelques caractères de la nature des corps célestes? Ces caractères ne peuvent-ils lui servir à marquer certains types auxquels les hypothèses astronomiques devront nécessairement se conformer? Ne devra-t-on pas, dès lors, déclarer irrecevable une combinaison de mouvements qui ne pourrait s'ajuster à aucun de ces types, lors même que cette combinaison sauverait les apparences?"—TRANSLATOR.

7. *Physics* 2.2.

This rather vague teaching does not allow us fully to grasp Aristotle's thought concerning the method of the astronomer and the method of the physicist. Really to penetrate his thought we must examine how he put this conception to work in his writings.

Eudoxus, Aristotle's predecessor by a few years, whose theories he studied assiduously, and Calippus, his contemporary and friend, had followed the method of the astronomer, exactly as defined by Plato. This method was, then, perfectly familiar to Aristotle. Yet he, for his part, followed another. Aristotle requires that the universe be a sphere, that the celestial spheres be hard, that each of them have a circular and uniform motion around the world's center, and that this center be occupied by the earth, an immobile earth. These were so many restrictive conditions that he imposed upon the hypotheses of astronomers, and he would not have hesitated to reject a combination of motions that presumed to dispense with any of them. Yet it was not because he considered them indispensable to saving the appearances registered by observers that he laid down these limiting conditions, but because according to him they alone were compatible with the perfection of the material of which the heavens are formed and with the nature of circular motion. While Eudoxus and Calippus, employing the method of the astronomer, controlled their hypotheses by examining whether or not they save the appearances, Aristotle wants to govern the choice of these hypotheses by propositions that are the outcome of certain speculations about the *nature* of heavenly bodies. His is the physicist's method.

What is the point of introducing this new method alongside the method of the astronomer, since it merely attempts to solve the astronomer's problem by another route? If the astronomer's procedure were capable of providing an altogether unambiguous answer to the question posed by Plato, one might well doubt that there was any gain. But if this is not how things stand, if it should turn out that the appearances can be saved by *various* combinations of circular and uniform motions, how then are we to choose from among these different, yet to the astronomer equally satisfactory, hypotheses? Must we in that case not appeal to the ruling of the physicist to make our selection, and would that not tend to show that the physicist's method is the indispensable complement to the method of the astronomer?

Now in point of fact the appearances *can* be saved by means of different combinations of circular and uniform motions, and the geometric acumen of the Greeks was far too developed for this truth to remain hidden from them for very long: Even very old astronomical systems, like that of Philolaus, for example, could only have germinated in minds thoroughly convinced of this principle: that the same relative motion can be obtained from different absolute motions.

In any case, one circumstance soon enforced an exceptionally clear realization of the truth that different hypotheses may render the phenomena equally well: This circumstance presented itself in the course of Hipparchus' investigations.

What Hipparchus proved was that the course of the sun can be represented either by supposing that this star describes a circle eccentric to the world, or by letting it be carried by an epicycle, provided the revolution of this epicycle is achieved in exactly the same time in which its center has completed a circle concentric with the world.

Hipparchus seems to have been very much struck by the agreement between the results of two such very different hypotheses. Adrastus of Aphrodisias, whose teachings have been preserved for us by Theon of Smyrna, records how Hipparchus felt about his own discovery:

> Hipparchus singled out as deserving the mathematician's attention the fact that one may try to account for phenomena by means of two hypotheses as different as that of eccentric circles and that which uses concentric circles bearing epicycles.[8]

Certainly, there is only *one* hypothesis that agrees with the nature of things (κατὰ φύσιν). *Every* astronomical hypothesis that saves the appearances is in harmony with this single hypothesis to the extent that the propositions entailed by it match the results of observation. This is what the Greeks meant when, speaking of different hypotheses which yielded the same resultant motion, they said that they agreed among themselves "accidentally" (κατὰ συμβεβηκὸς):

> It is obviously consistent with reason that there be agreement between the two mathematical hypotheses—the epicyclic and the eccentric—concerning the stellar movements. Both agree *accidentally* with *the one that conforms to the nature of things*, and this is what Hipparchus marveled at.[9]

8. Theon of Smyrna *Liber de Astronomia cum Sereni fragmento*, textum primus edidit, latine vertit, descriptionibus geometricis, dissertatione et notis illustravit T. H. Martin (Paris, 1849), chap. 26, p. 245; Theon of Smyrna "Exposition des connaissances mathématiques utiles pour la lecture de Platon." *Astronomie*, trans. J. Dupuis (Paris, 1892), pt. 3, chap. 26, p. 269.

9. Theon *Astronomia*, chap. 32 (Martin ed., p. 293; Dupuis ed., p. 299).

Which one of these different hypotheses, "accidentally" in agreement with each other, one saving the phenomena as well as the other and therefore, in the eyes of the astronomer, equivalent, conforms to nature? It is for the physicist to decide. If we are to believe Adrastus,[10] Hipparchus, more competent in astronomy than in physics, was incapable of making such a decision:

It is clear, for the reasons set forth, that, of the two hypotheses, each of which is a consequence of the other, the epicyclic appears to be the more common, more generally accepted, and better conformed to the nature of things. For the epicycle is a great circle of a rigid sphere, namely, that circle which the planet traces out as it moves on the sphere, whereas the eccentric is altogether different from the circle which conforms to nature, and it is traced out only "accidentally." Hipparchus, convinced that this is how the phenomena are brought about, adopted the epicyclic hypothesis as his own and says that it is likely that all the heavenly bodies are uniformly placed with respect to the center of the world and that they are united to it in a similar way. Not being sufficiently knowledgeable in physics, however, he did not distinguish properly between the *true* movement of the stars, which conforms to the nature of things, and their *accidental* movement, which is only an appearance. Nonetheless, in principle he holds that the epicycle of each planet moves along a concentric circle and that the planet moves along the epicycle.

By proving that two distinct hypotheses can agree "accidentally" and save the appearances of the solar movement equally well, Hipparchus greatly contributed to a more exact delimitation of the scope of astronomical theories. Adrastus set about proving that the eccentric hypothesis is entailed by the epicyclic;[11] Theon proved that the epicyclic hypothesis can, inversely, be considered a consequence of the eccentric hypothesis. These propositions, according to him, point up the impossibility of astronomy's ever discovering the *true* hypothesis, the one which conforms to the nature of things:

No matter which hypothesis is settled on, the appearances will be saved. For this reason we may dismiss as idle the discussions of the mathematicians, some of whom say that the planets are carried along eccentric circles only, while others claim that they are carried by epicycles, and still others that they move around the same center as the sphere of the fixed stars. We shall demonstrate that the planets "accidentally" describe each

10. Ibid., chap. 34 (Martin ed., p. 301; Dupuis ed., p. 303).
11. Ibid., chap. 26 (Martin ed., pp. 245–47; Dupuis ed., p. 269).

of these three kinds of circles—a circle around the center of the universe, an eccentric circle, and an epicyclic circle.[12]

If the decision that determines the true hypothesis escapes the competence of the astronomer, who attempts only to combine the abstract figures of the geometer and to compare them with the appearances reported by observers, it must then be reserved for the physicist, the man who has meditated on the nature of the heavenly bodies. He alone is competent to lay down the principles by means of which the astronomer will discern the one true hypothesis amidst the several suppositions that equally save the phenomena. This is precisely what the Stoic Posidonius asserted in his *Meteorology*. Geminus, in an abridged commentary on this work, reported Posidonius' doctrine; and Simplicius, for the purpose of clarifying Aristotle's comparison between the mathematician and the physicist, reproduces the passage from Geminus. It runs as follows:[13]

To physical theory (φυσικῆς θεωρίας) belongs the study of all that concerns the essence of the heavens and the stars, their power, their quality, their generation and destruction. And, by Zeus, physics also has the power of providing demonstrations concerning the size, shape, and arrangement of these bodies. Astronomy, on the other hand, is not prepared to say anything about the former. Its demonstrations concern the order of the heavenly bodies, taking it for granted that the heavens *are* truly ordered. Astronomy speaks of the shapes, sizes, and relative distances of the earth, the sun, and the moon. It speaks of eclipses, the conjunction of stars, the qualitative and quantitative properties of their movements. Now since astronomy depends on the study which considers figures in terms of quality, size, and number, it is quite right that it should require the assistance of arithmetic and geometry. And in dealing with these things, the only ones on which it is authorized to speak, astronomy must conform to arithmetic and geometry. It happens frequently that the astronomer and the physicist take up the same subject—for instance, they set out to prove that the sun is large or that the earth is round. But in such a case they do not proceed in the same way: The physicist must demonstrate every single one of his propositions by deriving it from the essence of bodies, or from

12. Ibid. (Martin ed., pp. 221–23; Dupuis ed., p. 251).

13. One might quarrel with Duhem's translation of certain phrases in this text. Nevertheless I have here, as throughout, translated Duhem's translation rather than the original, merely marking some of the phrases that seem off to me by inserting the Greek. The apparently minor point that φυσικῆς θεωρίας does not mean quite the same to Posidonius (or Simplicius) as *théorie physique* means to Duhem may turn out major: it signalizes, I think, that Duhem's analysis of the samenesses and differences in ancient and modern natural science needs rethinking, that "essential sameness" is come by too easily.—TRANSLATOR.

their power, or from what best accords with their perfection, or from their generation and their transformation. The astronomer, on the other hand, establishes his propositions by means of "what goes with" magnitudes and figures, or by means of the magnitude of the motion in question, or the time that corresponds to it. Often the physicist will fasten on the cause and direct his attention to the power that produces the effect he is studying, while the astronomer draws his proofs from circumstances externally related to that same effect. The astronomer is not equipped to contemplate causes, unable to tell us, for instance, what cause is responsible for the spherical shape of the earth and the stars. Sometimes, as for instance when he reasons about eclipses, he does not even try to lay hold of a cause. At other times he feels obliged to posit certain hypothetical modes of being which are such that, once conceded, the phenomena are saved (*καθ' ὑπόθεσιν εὑρίσκει τρόπους τινὰς ἀποδιδούς ὧν ὑπαρχόντων σωθήσεται τὰ φαινόμενα*). For example, the astronomer asks why the sun, the moon, and the other wandering stars seem to move irregularly. Now, whether one assumes that the circles described by the stars are eccentric or that each star is carried along by the revolution of an epicycle, on either supposition the apparent irregularity of their course is saved. The astronomer must therefore maintain that the appearances may be produced by either of these modes of being (*τρόπους*); consequently his practical study of the movement of the stars will conform to the explanation he has presupposed. This is the reason for Heraclides Ponticus' contention that one can save the apparent irregularity of the motion of the sun by assuming that the sun stays fixed and that the earth moves in a certain way. The knowledge of what is by nature at rest and what properties the things that move have is quite beyond the purview of the astronomer. He posits, hypothetically, that such and such bodies are immobile, certain others in motion, and then examines with what [additional] suppositions the celestial appearances agree. His principles, namely, that the movements of the stars are regular, uniform, and constant, he receives from the physicist. By means of these principles he then explains the revolutions of all the stars, both those which describe circles parallel to the equator and those which traverse circles that are at an angle to the equator.[14]

We have insisted on citing the passage in full because no other ancient text defines the respective roles of astronomer and physicist with equal precision. Posidonius, in order to drive home the astronomer's inability to grasp the true nature of the heavenly motions, appeals to the equivalence of the eccentric and epicyclic hypotheses discovered by Hipparchus; and side by side with this truth he mentions, citing Heraclides Ponticus, the equivalence of the geocentric and the heliocentric systems.

14. Simplicius *In Aristotelis physicorum libros quatuor priores commentaria* 2, ed. Diels (Berlin, 1882), pp. 291–92.

The Platonist Dercyllides, who lived at the time of Augustus, composed a word entitled: *Concerning the Spindles and Spindle Whorls Mentioned in Plato's Republic* (Περὶ τοῦ ἀτράκτου καὶ τῶν σφονδύλων ἐν τῇ πολιτείᾳ παρὰ Πλάτωνι λεγομένων). It contained astronomical theories of which Theon of Smyrna has preserved a summary for us.[15]

The Platonist Dercyllides, it turns out, conceived of the relations between astronomy and physics exactly as did the Stoic Posidonius:

> Just as in geometry and music it is impossible to deduce what follows from the principles unless one lays down hypotheses, so in astronomy one must first exhibit the hypotheses from which the theory of the motion belonging to the wandering stars derives. But perhaps one should, before all else, lay down the principles on which the study of mathematics rests, principles conceded by everyone.[16]

Posidonius had said that the investigation of what is at rest and what in motion belongs to the physicist. So Dercyllides is careful to rank the propositions determining what bodies are absolutely at rest among principles which precede the hypotheses of astronomy:

> Since it does not conform to reason that all bodies should be in motion, nor that all should be at rest, but some move and others are immobile, one must find out what in the universe is necessarily at rest and what in motion.

He adds that one must believe that the earth, which is the hearth in the house of the gods according to Plato, remains at rest and that the planets move, along with the entire celestial vault that contains them.

Dercyllides does not leave the mathematician with the option of disregarding the principles established and formulated by the physicist. The mathematician has no right to advance hypotheses that contradict physical principles. The assumption that Posidonius and Geminus attribute to Heraclides Ponticus, namely, that the earth might be in motion and the sun at rest, would be such a violation of the physicist's principles. Dercyllides "rejects with horror those who, by stopping the bodies that move and setting in motion those that are immobile by virtue of their nature and their place, overturn the foundations of mathematics."

Among the principles so severely imposed upon the astronomer's deference, Dercyllides does not include the necessity of reducing all heavenly movements to revolutions around the world's center. Planetary motion

15. Theon *Astronomia*, chaps. 39, 40–43.
16. Ibid., chap. 41 (Martin ed., p. 327; Dupuis ed., p. 323).

along an epicycle whose own center describes an orbit concentric with the world does not to him, seem a violation of sound physics. As Theon of Smyrna reports:[17]

He [Dercyllides] does not think that eccentric circles are the cause of the movement that makes the distance between a planet and the earth vary. He thinks that every moving thing in the sky is carried around a single center of both motion and world. [He holds, therefore, that the eccentric movements] exhibited by the planets are not their "principal" movements but "accidental" ones. Such movements are, as we demonstrated earlier, resultants of the compounding of epicyclic and concentric movements, which are traced out within the thickness of an orbital shell that is homocentric with the world. For every sphere has two surfaces, an interior surface, which is concave, and an exterior surface, which is convex. Between these two surfaces the planet moves in an epicycle and in a concentric circle. The effect of this movement is to describe "accidentally" an eccentric circle.

Why should Dercyllides consider planetary motion along a circle eccentric to the world contrary to the principles of his physics? And why, on the other hand, does this same physics permit a planet to describe an epicycle whose center traverses a circle concentric with the world? No explicit answer to the question is furnished by Theon of Smyrna's report on Decyllides' doctrines. But we may assume that the reasons invoked by Dercyllides to justify his preference are no different from those which induced Adrastus of Aphrodisias to adopt a very similar position.

According to the testimony of Theon of Smyrna,[18] Adrastus of Aphrodisias ascribed to every planet an orbital shell contained by two spherical surfaces concentric with the universe. Within the shell is a full sphere occupying its entire thickness. The planet is then set into this full sphere. The orbital shell carries the full sphere in its revolution around the center of the world, while the full sphere turns upon its own axis. By means of this mechanism the planet describes an epicycle whose center traverses a circle concentric with the world.[19]

17. Ibid. (Martin ed., p. 331; Dupuis ed., p. 325).

18. Ibid., chaps. 31, 32 (Martin ed., pp. 275, 281–85; Dupuis ed., pp. 289, 293–95).

19. The French is quite obscure: It reads as follows: "Au témoignage de Théon de Smyrne, Adraste d'Aphrodisie attribue à chaque astre errant un orbe que contiennent deux surfaces sphériques concentriques à l'Univers. A l'intérieur de cet orbe se troue une sphère pleine qui en occupe toute l'épaisseur. L'astre, enfin, est enchâssé en cette sphère pleine. L'orbite entraîne la sphère pleine en la rotation qu'elle effectue autour du centre du Monde, tandis que la sphère pleine tourne sur elle-même. Par ce mécanisme, la planète décrit un épicycle dont le centre parcourt

Adrastus of Aphrodisias, and Theon of Smyrna after him, held that this mechanism conforms to the principles of sound physics. For them these principles are, then, no longer what they were for Aristotle. Physical principles seem on their view to reduce to the following single proposition: The heavenly movements should be represented by a combination of rigid spheres, whether hollow or full, each of which turns with a uniform rotation about its own center.

What nature requires is this: those circular and helical lines should not be traced out by the stars themselves moving of themselves[20] in a direction contrary to the movement of the world; and the movement of the stars should not be explained by literally tying them to circles each of which moves around its own particular center and carries the attached star with it. After all, how could such bodies be tied to incorporeal circles?

Eudoxus and Calippus had accounted for the movement of the heavens by appealing to the agency of a variety of rigid spheres. The Stoic Cleanthes, rejecting their account,[21] had claimed that each star is self-propelled, *itself* describing the geometric curve called the "hippopede," which Eudoxus and Calippus had obtained indirectly by compounding the rotations of several spheres. It was against the view of Cleanthes that Dercyllides was arguing: According to him the "hippopede" should be understood as a line that is traced out only "accidentally," since no movement except the uniform rotation of rigid spheres is "natural" for the heavens.

This doctrine of Dercyllides was obviously the one which inspired Adrastus of Aphrodisias and Theon of Smyrna. They applied it, no doubt here too following Dercyllides, not only to hippopedic motion but also to eccentric and epicyclic motion. They rejected every theory which confined itself to a purely geometric description of the planetary paths. They accepted the theory which made the planets describe epicycles whose centers traverse a circle concentric with the world, because they had discovered a procedure which allowed for the imposition of such a trajectory upon a

---

un cercle concentrique au Monde." Note the switch from "orbe" to "orbite," which causes part of the trouble. For some, though insufficient, enlightenment, see Edward Rosen's Introduction to his *Three Copernican Treatises: the Commentariolus of Copernicus; the Letter against Werner; the Narratio prima of Rheticus* (New York: Dover Publications, Inc., 1959), especially pp. 18ff.—TRANSLATOR.

20. According to Martin (p. 274, n. 5), the manuscript reads: *τὰ ἄστρα αὐτὰ κατὰτ' αυτὰ*; Martin emends this to *κατὰ ταῦτα*; and Dupuis, in our opinion mistakenly, accepts the emendation.

21. Joannes Stobaeus *Eclogarum physicarum et ethicarum libri duo*, bk. 1, "Physica," chap. 25 (ed. Augustus Meineke; Leipzig, 1860), vol. 1, p. 145.

planet by having rigid spheres, suitably arranged, turn upon themselves. To Adrastus and Theon a hypothesis appeared compatible with the nature of things if a competent craftsman could embody it in metal or wood. Many are those, even today, who have hardly a different notion of sound physics.

Theon, moreover, roundly admits the great importance he attaches to such material models. He reports that he constructed an orrery that could serve as a model of Plato's astronomical theory:

> For Plato says that we would be engaging in futile labor if we tried to explain these phenomena without images that speak to the eyes.[22]

He goes so far as to attribute the theory which rejects eccentric movements in favor of movements in an epicycle whose center traverses a circle eccentric with the world to Plato himself![23]

Actually, Plato was never obliged to state his preference on this point, since neither the eccentric nor the epicyclic hypothesis had ever occurred to him. Revolutions homocentric with the world are the only ones to which he ever made allusion in his writings, as Proclus quite correctly asserts in several places.[24]

Nevertheless, Adrastus and Theon were not entirely wrong in claiming that they were appealing to the principles of Platonic physics. Plato had ascribed a rotary motion around its own center to each and every star. So it seems that the rotation of an epicyclic sphere around itself would not in any way have violated his teachings concerning the heavenly movements. He might even have adopted the theory of the sun proposed by Hipparchus. Only Aristotle's physics was truly incompatible with the existence of epicycles: Incapable of any alteration, inaccessible to all "violence," a heavenly essence could not, according to Aristotle's physics, manifest any but its own "natural" movement, and its only natural movement was a uniform rotation around the center of the universe.

Adrastus of Aphrodisias and Theon of Smyrna, probably Dercyllides as well, require that the mathematician so choose his astronomical hypotheses that they conform to the nature of things. But this conformity is no longer gauged by the principles of physics Aristotle had laid down. A

22. Theon *Astronomia*, chap. 16 (Martin ed., p. 203; Dupuis ed., p. 239).
23. Ibid., chap. 34 (Martin ed., p. 303; Dupuis ed., p. 305).
24. Proclus Diadochus *In Platonis Timaeum commentaria*, ed. Diehl (Leipzig, 1903–6), (Tim. 36 D), (Tim. 39 D, E), (Tim. 40 C, D), vol. 2, p. 364; vol. 3, pp. 96, 146 respectively.

hypothesis' conformity to nature is now judged in terms of the possibility of constructing a mechanism of suitably fitted rigid spheres, which can be taken to represent the movements of the heavens. Planetary motion resulting from the revolution of an eccentric whose center traverses a circle concentric with the world can be "modeled" by the "turner." Such an hypothesis is, therefore, acceptable to the physicist, though it violates the nature of the Peripatetic "quintessence." It is just as acceptable as the system of homocentric spheres of Eudoxus, Calippus, and Aristotle.

Progress in astronomy would quickly make the position taken by Adrastus and Theon untenable. The moment Ptolemy, attempting to represent the irregularities of planetary motion, had each planet borne on an epicycle whose center, instead of maintaining a constantly equal distance from the center of the universe, described a circle eccentric to the world, the orrery imagined by Adrastus of Aphrodisias and Theon of Smyrna became incapable of representing the celestial movements. And with each complication of Hipparchus' primitive hypotheses that Ptolemy had to add in order to save the phenomena, the impotence of Adrastus' and Theon's spheres became more apparent. Certainly, no Peripatetic could regard the hypotheses of Ptolemy's *Syntaxis* as conforming to the principles of physics, since these hypotheses do not reduce all heavenly movements to homocentric revolutions. But neither could a disciple of Adrastus or Theon have looked upon them as physically acceptable, for, surely, no craftsman could have constructed a wooden or metal representation of them. Ptolemy's followers were bound—on pain of abandoning their own doctrines—to liberate astronomical hypotheses from the conditions to which physicists had generally subjected them.

Ptolemy assigned every planet an orbital shell of a certain thickness which was contiguous with the shells of the planets that precede or follow it.[25] The planet moved between the spherical surfaces of this shell, which was concentric with the world and which delimited the planet's orbit. This movement was explained in the *Syntaxis* in terms of a large number of very complicated hypotheses. How, exactly, are we to understand the relation between these hypotheses and the principles of physics? Or, to put it differently, what restrictive conditions can physics still impose on the hypotheses of astronomy? This is a question over which Ptolemy, more geometer and astronomer than philosopher, does not linger. Still, he does touch on it,[26] in a passage whose purport becomes remark-

25. *Almagest* 9.1 (Halma ed., vol. 2, pp. 113–15).
26. Ibid., 13.2 (pp. 374–75).

ably clear and forthright when it is construed in the light of everything that has so far been said.

The astronomer, intent on finding hypotheses that will do the job of saving the apparent movements of the stars, has no guide except the rule of maximum simplicity:

> We must, as best we can, adapt the simplest hypotheses to the heavenly movements. But if these prove insufficient, we must select others that fit better.

The accurate representation of the heavenly movements may well oblige the astronomer gradually to complicate his assumptions. But the complexity of the system at which he arrives in this way cannot be a reason for rejecting it if it fits accurately with observation:

> If every apparent movement gets saved, as warranted by the hypotheses, why should anyone find it surprising that it is from such complicated motions that the movements of the heavenly bodies result?[27]
>
> Let no one judge the real difficulties of the hypotheses in terms of the constructions we have devised. It is not fitting to compare things human with things divine. We should not base our trust in things so high on examples drawn from what is most greatly removed from them: For is there anything that differs more from changeless beings than beings that are constantly changing? Or is there anything that differs more from beings which are interfered with by the entire universe than the beings which do not even interfere with themselves?

Foolish, then, the desire to impose on the movements of the heavenly bodies the obligation of letting themselves be modeled by wooden or metal contraptions:

> So long as we attend to these models, which we have put together, we find the composition and succession of the various motions awkward. To set them up in such a way that each motion can freely be accomplished hardly seems feasible. But when we study what happens in the sky, we are not at all disturbed by such a mixture of motions.

Certainly, Ptolemy means to indicate in this passage that the many motions he compounds in the *Syntaxis* to determine the trajectory of a planet have no physical reality; only the resultant motion is actually produced in the heavens.

Among the movements that the astronomer is thus led to assign to the planets because he seeks to save the phenomena, might he come upon

27. Συμβεβηκέναι—to come about "by accident" (κατά συμβεβηκὸs); in modern terms, a motion resulting from the composition of other motions.

any which are repugnant to the nature of the celestial essence? None whatever:

In the region where these movements occur there is no essence naturally endowed with the power to oppose such movements. What is found there yields with indifference to the natural movement of every planet and allows it to pass even though the several movements occur in different directions. So all the stars can pass and all can be perceived through the fluids which are there homogeneously spread out.

In spite of the brevity of this statement, it gives us a clear idea of Ptolemy's teachings on astronomical hypotheses.

The various revolutions—in concentric or eccentric circles or in epicycles—which we put together to obtain the paths of the planets are contrivances assembled with a view to saving the phenomena by means of the simplest hypotheses we can find. We must be very much on our guard against the idea that these mechanical constructions have the least reality up in the sky. The orbital shells of the planets are filled with a fluid that offers no resistance to the movements of the bodies immersed in it. Surrounded by this fluid, the planets trace out their more or less complicated trajectory, without any rigid spheres to guide them along their course. Ptolemy's astronomical teachings are, of course, more sophisticated, but he depends on a physics quite similar to that of Cleanthes. He simply ignored the objections that Dercyllides, Adrastus of Aphrodisias, and Theon of Smyrna had earlier made against this physics.

Ptolemy's attitude toward the theorem of Hipparchus makes it very clear that he has broken with the principles to which Adrastus and Theon had appealed. The sun's movement is equally well saved whether one has it describe a circle eccentric to the world or lets it turn in an epicycle whose center remains always at the same distance from the center of the universe. Which of these hypotheses is the one that a sound physics requires us to adopt? According to Adrastus and Theon, it is the epicyclic hypothesis, because a model constructed of rigid spheres, one encased within the other, would then allow us to represent the sun's movement. According to Ptolemy, "it is more reasonable to select the epicyclic hypothesis, since it is more simple and since it assumes only a single movement and not two."[28]

The doctrine Ptolemy expounds in this passage seems to have been adopted without reservation by Proclus, who deals with it in various parts of his writings.

28. *Almagest* 3.4 (Halma ed., vol. 1, pp. 183–84).

In particular, he examines it toward the end of the book in which, under the title of *Hypotyposes*, he presents a list of Ptolemy's astronomical hypotheses.[29]

Proclus' entire effort is geared to showing that the hypothetical eccentric and epicyclic motions by the compounding of which the movements of the planets are reproduced are pure abstractions. These motions exist nowhere but in the mind of the astronomer. They are nothing in the heavens. The only movement that is real is the complex, undecomposed movement of each planet.

This assertion runs directly counter to the doctrine according to which the heavenly bodies can, because of their essence, undergo only circular and uniform movements. Proclus was well aware of this and said so:

The astronomers who presupposed the uniformity of the movements of the heavenly bodies did not realize that the essence of these movements is, on the contrary, irregularity.

By virtue of the principle laid down by their physics, these astronomers looked upon the complicated and irregular movement of a planet, the movement manifest to sight, as the resultant of several simple motions effected along an eccentric and an epicycle. For them the latter were the only real movements, the former was a "mere appearance."

Now as regards these eccentrics and epicycles, there are two prevailing opinions:

Either these circles are merely fictive and ideal; or they have a real existence amidst the planetary spheres and are to be found inside these spheres.

If the eccentrics and epicycles, or rather the planetary movements that describe them, are purey conceptual, why should they be the only real and genuine movements and the observed movements "mere appearances"? These who maintain this doctrine

. . . forget that these circles exist only in thought; they interchange natural bodies and mathematical concepts; they account for natural movements by means of things which have no existence in nature.[30]

29. Ptolemy *Hypothèses et époque des planètes* and Proclus Diadochus *Hypotyposes ou représentations des hypothèses astronomiques*, translated for the first time from Greek into French by Halma (Paris, 1820); Proclus *Hypotyposes* (Halma ed., pp. 150–51).

30. The text has *ἐκ τῶν ἐοἰκούντωνἐν τῇ φύσει*, obviously, it should read: *οὐκ οἰκούντων*.

On the other hand, if one opts for the alternative and maintains that the eccentrics and epicycles are not conceptual but bodies physically present in the celestial essence, one soon runs into contradictions:

For by conceding that the planets' irregular movements are really produced by these circles and that the latter really exist on the vault of the heavens, these astronomers destroy the continuity of the spheres which contain these circles and move them, for some of them move in this, others in the opposite direction, and the former follow a different law than do the latter.

Since, however, the combinations of movements proposed by astronomers are purely conceptual and devoid of reality, there is no need to justify them by means of physical principles. They need only be arranged in such a way that the appearances are saved. Astronomers

do not arrive at conclusions by starting from hypotheses, as is done in the other sciences; rather, taking the conclusions as their point of departure, they strive to construct hypotheses from which effects conformable to the original conclusions follow with necessity (*οὐκ ἀπὸ τῶν ὑποθέσεων τὰ ἑξῆς συμπεραίνουσιν, ὥσπερ αἱ ἄλλαι ἐπιστῆμαι, ἀλλ' ἀπὸ τῶν συμπερασμάτων τὰς ὑποθέσεις ἐξ ὧν ταῦτα δεικνύναι ἔδει πλάττειν ἐγχειροῦσι*).

When these hypotheses have enabled us to decompose the complex movements of the planets into simpler ones, we should not think that we have now come upon the real movements that lie behind the apparent ones. The real movements *are* the apparent movements. The end achieved is more modest: we have simply made the celestial phenomena accessible to calculation:

These hypotheses are framed with an eye to discovering the form of the movements of the planets, which actually move in conformity with what appears. But thanks to the hypotheses, we can start to measure the details of the planetary appearances (*ἵνα γένηται καταληπτὸν τὸ μέτρον τῶν ἐν αὐτοῖς*).

Earlier, Ptolemy had warned astronomers against the temptation of comparing divine with human things. This call to modesty—so becoming to human science—was heeded by Proclus,[31] with whose Platonism it was in perfect harmony.

Because of our weakness, imprecision gets introduced into the series of images by which we represent what is. To know, *we* must use imagination, sense, and a multitude of other instruments, because the gods

31. Proclus *In Platonis Timaeum commentaria* B (Tim. 29 C, D), (Diehl ed., vol. 1, pp. 352–53).

have reserved these things for the One among them, the divine Mind. When we are dealing with sublunary things, we are content, because of the instability of the material which goes to constitute them, to grasp what happens in most instances. But when we want to know heavenly things, we use sensibility and call upon all sorts of contrivances quite removed from likelihood. As a result, when any of these things is the subject of investigation, we, who dwell, as the saying goes, at the lowest level of the universe, must be satisfied with "the approximate" (τὸ ἐγγὺς). That this is the way things stand is plainly shown by the discoveries made about these heavenly things—from different hypotheses we draw the same conclusions relative to the same objects. Among these hypotheses there are some which save the phenomena by means of epicycles, others which do so by means of eccentrics, still others which save the phenomena by means of counterturning spheres devoid of planets.[32]

Surely, the gods' judgment is more certain. But as for us, we must be satisfied to "come close" to those things, for we are men, who speak according to what is likely, and whose lectures resemble fables.[33]

Astronomy cannot grasp the essence of heavenly things. It merely gives us an image of them. And even this image is far from exact: it merely comes close. Astronomy rests with "the nearly so." The geometric contrivances we use to save the phenomena are neither true nor likely. They are purely conceptual, and any effort to reify them must engender contradictions. Combined for the sole purpose of furnishing conclusions that conform to observation, they are by no means determined unambiguously. Very different hypotheses may yield identical conclusions, one saving the appearances as well as the other. Nor should we be surprised that astronomy has this character: It shows us that man's knowledge is limited and relative, that human science cannot vie with divine science. Such is Proclus' teaching, surely, a far cry from the ambitious physics of Aristotle's *On the Heavens* and *Metaphysics*, the physics that claims to have carried speculation on the essence of heavenly things so far as to have arrived at the fundamental principles of astronomy.

In more than one respect, Proclus' doctrine can be likened to positivism. In the study of nature it separates, as does positivism, the objects accessible to human knowledge from those that are essentially unknowable to man. But the line of demarcation is not the same for Proclus as it is for John Stuart Mill.

32. Proclus is here referring to the ἀναλίττουσαι σφαῖραι of Eudoxus, Calippus, and Aristotle.

33. Proclus is obviously alluding to *Timaeus* 29 and elsewhere. It is very curious that Duhem does not in fact do what he tells us we must—"go back to Plato."—TRANSLATOR.

The study of the elements and compounds that form the sublunary world Proclus gives over to human reason; their nature we can know; we can construct a physics of bodies subject to generation and corruption. Of the heavenly substances, however, we can know only appearances; their *nature* is understandable only to the divine *logos.*

The moment the same nature was ascribed to both heavenly and sublunary bodies, this doctrine had to be modified. By extending to all bodies what Proclus had reserved for the stars, by declaring that only the phenomenal effects of any material are accessible to human knowledge whereas the inner nature of this material eludes our understanding, modern positivism came into being.

Simplicius, eclectic, and without inclination for extreme solutions, takes up a sort of mean position between Aristotle and Proclus.

With Aristotle, he holds that circular and uniform movement is the essential movement of the heavenly bodies; he merely refuses to go along with the Stagirite's thesis that every portion of the "quintessence" necessarily revolves around the center of the universe. The irregular movements of the planets are not, as Proclus had claimed, their only real movements. Rather, they are the complicated appearances produced by the combined action of several circular and uniform movements.

These principles, formulated in physics, set astronomy the following problem: to decompose the movement of each planet into circular and uniform motions. But once it has assigned this task, the study of the heavenly essence does not provide the astronomer with the means of completing it: it does not inform him which movements are the genuine circular and uniform movements, the ones which really underlie the apparent course of a planet.

So the astronomer takes up the question in a different way. He imagines certain circular and uniform movements produced either by homocentric spheres devoid of planets or by eccentrics and epicycles. He combines such movements until he succeeds in saving the phenomena. Once this object is obtained, he should, however, be very careful not to jump to the conclusion that his hypotheses represent the planets' real movements. The simple movements that he has imagined and combined are no more the real movements of the heavenly bodies than are the irregular and complicated movements which are obvious to our senses.

Since the astronomer's hypotheses are not realities but merely fictions, the whole purpose of which is to save appearances, we should not be surprised that different astronomers attempt to achieve this purpose by means of different hypotheses.

Such, we believe, is Simplicius' doctrine, as it is, in our opinion, clearly stated in various passages of his writings. Here are some of the passages we have in mind:

Surely, the fact that opinions differ on these hypotheses is no ground for indictment: For the object is to find out by means of what hypothesis we shall be able to save the phenomena. There is no reason, then, to wonder when different astronomers endeavor to save phenomena by starting out from different hypotheses (*Δῆλον δὲ, ὅτι τὸ περὶ τὰς ὑποθέσεις ταύτας διαφέρεσθαι οὐκ ἔστιν ἔγκλημα τὸ γὰρ προκείμενόν ἐστι, τίνος ὑποτεθέντος σωθείη ἂν τὰ φαινόμενα; οὐδὲν οὖν θαυμαστὸν, εἰ ἄλλοι ἐξ ἄλλων ὑποθέσεων ἐπειράθησαν διασῶσαι τὰ φαινόμενα*).[34]

The curious problem of astronomers is the following: First, they provide themselves with certain hypotheses: The ancients, the contemporaries of Eudoxus and Calippus, adopted the hypothesis of "counterturning spheres"; Aristotle, who in his *Metaphysics* teaches the system of spheres, must be counted among them. The astronomers who followed proposed the hypothesis of eccentrics and epicycles. Starting from such hypotheses, astronomers then try to show that all the heavenly bodies have a circular and uniform motion, that the irregularities which become manifest when we observe these bodies—their now faster, now slower motion; their moving now forward, now backward; their latitude now southern, now northern; their various stops in one region of the sky; their at one time seemingly greater, and at another time seemingly smaller diameter—that all these things and all things analogous are but appearances and not realities. . . .[35]

To save these irregularities, astronomers imagine that each star is moved by several motions at the same time—some assuming movements along eccentrics and epicycles, others appealing to spheres homocentric with the world (the so-called counterturning spheres). But just as the stops and the retrograde motions of the planets are, appearances notwithstanding, not viewed as realities (they are no more real than the numerical additions and subtractions with which we meet in studying these motions), so an explanation which conforms to the facts does not imply that the hypotheses are real and exist. By reasoning about the nature of the heavenly movements, astronomers were able to show that these movements are free from all irregularity, that they are uniform, circular, and always in the same direction. But they have been unable to establish in what sense, exactly, the consequences entailed by these arrangements are merely fictive and not real at all. So they are satisfied to assert that it is possible, by means of circular and uniform movements, always in the same direction, to save the apparent movements of the wandering stars.[36]

34. Simplicius *In Aristotelis quatour libros de Coelo commentaria* 1. 6 (Karsten ed., p. 17, col. b; Heiberg ed., p. 32).
35. Ibid. 2.28 (Karsten ed., p. 289, col. b; Heiberg ed., p. 422).
36. Ibid. 2.44 (Karsten ed., p. 219, col. a; Heiberg ed., p. 488).

This doctrine is in every point like the one Posidonius had formulated and of which Geminus has preserved a report for us. It is therefore not surprising that Simplicius inserted it into his commentary on Aristotle's *Physics* and that he seems to have regarded it as the best definition of the respective roles of the mathematician and the physicist.

2
Arabic and Jewish Philosophy

The Greeks had persevered in applying their talent for geometry to the task of analyzing the complex and irregular movements of the wandering stars into a small number of simple circular motions; and their exertion was matched by their success. Soon they turned their equally great talent for logic and metaphysics to the study of the kinetic combinations that had been devised by the astronomers. And after some initial hesitation they balked at the idea that the eccentrics and epicycles are bodies, really up there on the vault of the heavens. For the Greeks they were simply geometric fictions requisite to the subjection of celestial phenomena to calculation. If these calculations are in accord with the results of observation, if the "hypotheses" succeed in "saving the phenomena," the astronomer's problem is solved. Astronomical hypotheses are, in this sense, useful. But only the physicist would be authorized to say whether or not they conform to reality. Generally speaking, the principles *he* is able to affirm are too general, too remote from particulars, to empower him to pronounce that kind of judgment.

The prodigious geometric ingenuity of the Greeks did not form part of the heritage they passed on to the Arabs. Nor did the Arabs have the Greeks' remarkably sure and precise logical sense. They brought only some very minor improvements to the hypotheses whereby the Greek astronomers had managed to resolve the complex course of the planets into simple motions. Moreover, when they did at last come to examine these hypotheses in an attempt to make out their nature, their vision could not match the penetration of a Posidonius, a Ptolemy, a Proclus, or a Simplicius; slaves to their imagination, they tried to see and touch what the Greek thinkers had declared fictive and abstract. They wanted to embody, in rigid spheres rolling about on the vault of the heavens, the eccentrics and epicycles that Ptolemy and his successors had proffered as contrivances of calculation.

Furthermore, it was only quite late, we find, that the Arabic astronomers came to feel the need to question the hypotheses of astronomy. For a long time, those who had studied the *Almagest* confined themselves to composing paraphrases, summaries, and commentaries, and to constructing tables facilitating application of its principles. In no way did they inquire into the purport and nature of the assumptions by which the entire Ptolemaic system is supported. In the writings of Abul Wefa, of al-Fergani, and al-Battani one would search in vain for the least insight into the degree of reality that is to be ascribed to the eccentrics and epicycles. Science was, in other words, passing through one of those periods when its experts are wholly given over to a concern for perfecting applications of theory and methods of observation and have neither the leisure nor the desire to question the soundness of the foundations of the edifice of science. In the course of its development, science has gone through several such periods during which the critical faculty drowses; but soon thereafter it wakes afresh, more eager now to query the principles of physical doctrine than to deduce new conclusions from it.

To come upon an author who discusses the nature of the mechanisms conceived by Ptolemy, we have to skip all the way to the end of the ninth century.

At that time the learned and productive Sabian astronomer Thabit ibn-Qurra composed a treatise in which he tried to assign a physical constitution to the heavens that might square with the Ptolemaic system. This treatise is not among the works of the author that have come down to us in a Latin translation; we know of it only through the testimony of Maimonides and Albert the Great, who had direct access to it. From them we learn that Thabit ibn-Qurra constructed the heavens by means of rigid orbital shells—some hollow, some filled—that rolled about in a fluid ether that could condense or dilate.

The same bias which led Thabit ibn-Qurra to "realize" the Ptolemaic hypotheses materially, to divest them of their purely geometric, abstract, character by "incarnating" them in rigid or yielding bodies, continued to dominate the scientific efforts of the Moslem thinkers. More than a century after the death of Thabit we find it giving orientation to the investigations of Ibn al-Haitam (author of the work on optics which, under the name Perspective of al-Hazan, was so much in vogue until the Renaissance).

Ibn al-Haitam's *Epitome of Astronomy*, written in Arabic, was translated into Hebrew by Jacob ben Machir (Prophatius) and then translated from the Hebrew into Latin by Abraham of Balmes. In passing through these two successive versions, the Arabic preamble to this treatise was transformed into an utterly amazing mish-mash. Amidst the innumerable bits of nonsense with which this *proemium*[1] is adorned, there are nevertheless a few passages on the verge of intelligibility, the original thought of the author managing, despite everything, to shine through. Here we find the Arabic astronomer up in arms against those who, in order to account for the celestial motions,

> construct abstract demonstrations by means of an ideal point's motion along the circumference of fictive circles. . . . Such demonstrations make sense only in terms of the object these authors had in mind, namely, the measuring they had defined and described. . . . The circles and the fictional point whose motion Ptolemy viewed entirely abstractly we shall place on spherical or plane surfaces animated by the same motions. We thus obtain a representation that is both more exact and clear to the understanding. . . . Our demonstrations will be shorter than the ones which use only this ideal point and these fictive circles. . . . We have studied the various motions that occur within the orbs in such a way that for each we provided a corresponding simple, continuous, and eternal motion of a spherical body. The bodies we thus assign to the motions can all go into action at the same time without this action's being in conflict with the position that has been assgned to them: they will not run into anything that might knock against them and dent or shatter them. Furthermore, while moving, these bodies will remain continuous with the intervening substance. . . .

Thabit ibn-Qurra and Ibn al-Haitam belong to the same intellectual family to which Adrastus of Aphrodisias and Theon of Smyrna belong. Abstract hypotheses reduced to geometric fictions cannot satisfy them, no matter how well they may suit the phenomena. But once they have succeeded in representing these hypotheses by means of bodies that a potter or a sculptor might fashion, bodies so arranged that they can be made to revolve around one another, their imagination, its needs now gratified, mistakes itself for reason and thinks that it has penetrated the very nature of things.

We meet such minds in every age. They turn up long after Ibn al-

1. Maurice Steinschneider, "Notice sur un ouvrage astronomique inédit d'Ibn Haïtam," *Bulletino di bibliografia e di storia delle scienze matematiche e fisiche* (B. Boncompagni, 1883), vol. 14, pp. 733–36.

Haitam. In the preface to his translation of the *Epitome of Astronomy*,[2] Prophatius tells us that a man "who came from a distant country had found the demonstrations in the book of al-Fergani out of joint with the nature of existing things, and urged him to translate" the work of Ibn al-Haitam. Now the machinery of rigid spheres proposed by al-Haitam—really a development of an idea of Simplicius—suggested mechanical models of the Ptolemaic system and, because of this, the *Summary of Astronomy* greatly contributed to the eventual triumph of the Ptolemaic system among the Christians of the West. Nevertheless, it would not take long before the hypotheses developed in Ibn al-Haitam's treatise would be attacked in the name of the principles of physics, precisely because *these* hypotheses *were* claimed to represent the nature of things.

To recapitulate, the hypotheses of astronomy can be viewed as mathematical fictions which the geometer combines for the purpose of making the celestial motions accessible to his calculations; or they can be viewed as a description of concrete bodies and of movements that are actually realized. In the first case, only one condition is imposed on hypotheses, namely, that they save the appearances; in the second, the intellectual freedom of the astronomer turns out to be much more limited, for if he is an advocate of a philosophy which claims to know something about the celestial essence, he will have to reconcile his hypotheses with the teachings of that philosophy.

Ptolemy and the Greek thinkers who came after him adopted the first of these two opinions. They were therefore able to put their geometric theories together without concern for the various physical opinions over which they argued among themselves or with their contemporaries. They could select their assumptions without troubling about anything except agreement between the results of their calculation and the facts of observation.

The Arabic astronomers, however, following Thabit ibn-Qurra and Ibn al-Haitam, wanted their hypotheses to correspond to the true movements of really existing hard or yielding bodies. So their hypotheses were accountable to the laws of physics.

Now the physics endorsed by the majority of Islamic philosophers was the Peripatetic physics, the philosophy which Sosigenes and Xenarchus had long ago opposed to the astronomy of eccentrics and epicycles by showing that the reality of the latter cannot be reconciled with the truth

2. Ibid., p. 723.

of the former. The realism of the Arabic philosophers could not but incite the Islamic Peripatetics to passionate and merciless battle against the doctrines of the *Almagest.*

That battle was to last throughout the twelfth century.

Maimonides tells us that ibn-Badia (the Avempace of the Latin Scholastics), rejected the epicycles as incompatible with the principles of Aristotle's physics. According to Averroes and al-Bitrogi, Abu Beker ibn-Tofail (the Abu Bacer of the "Schools") went still further: He tried to construct an astronomy from which epicycles and eccentrics were both banished.

Averroes' debt to the philosophers who resisted the hypotheses of the Almagest was particularly great: "His philosophy takes off directly from ibn-Badia; ibn-Tofail was the master of his fate."[3] His intellectual formation predisposed him, therefore, to join in the battle against Ptolemy.

To this he was no less inclined by his fanatical devotion to Aristotle. Aristotle, he says in the preface to his commentary on the *Physics,*

> founded and completed logic, physics, and metaphysics. I say that he founded them because the works written before him on these sciences are not worth talking about and are quite eclipsed by his own writings. And I say that he completed them because no one who has come after him up to our own time, that is, for nearly fifteen hundred years, has been able to add anything to his writings or to find any error of any importance in them.

The author of these lines could not but regard as erroneous the assumptions that Hipparchus and Ptolemy had substituted for the principles advanced in the *On the Heavens.*

Averroes' commentary on the *On the Heavens* does more than explain the system of homocentric spheres and supply all the support for this system that Aristotelian physics can provide. In also contains a very solid and penetrating critique of the system developed in the *Almagest.*[4] He returns to this critique when he comments on Book XII of the *Metaphysics.*[5]

We cannot here go into Averroes' long argument against the hypotheses of Ptolemy. We have to confine ourselves to culling those passages

3. Ernest Renan, *Averroès et l'Averroïsme, essai historique* (Paris, 1852), p. 11.

4. *Aristotelis De Caelo cum Averrois Cordubensis commentariis,* lib. 2, summae secundae quaes. 2, comm. 32; lib. 2, quaes. 5, comm. 35.

5. *Aristotelis Metaphysica cum Averrois Cordubensis expositione,* lib. 12, summae secundae cap. 4, comm. 45.

where the Commentator reveals how he thinks about astronomical theories in general.

One of these passages is quite remarkable:

> We find nothing in the matematical sciences that would lead us to believe that eccentrics and epicycles exist.
>
> For astronomers propose the existence of these orbits as if they were principles and then deduce conclusions from them which are exactly what the senses can ascertain. In no way do they demonstrate by such results that the assumptions they have employed as principles are, conversely, necessities.[6]
>
> Now, through logic we know that every demonstration goes from what is better known to what is more obscure. If what is better known is posterior to what is less known, we have a "demonstration *quia*." But if what is better known precedes what is less known, two situations may arise: It may be that the existence of the object of demonstration is obscure, while its cause is known. In such a case we have an absolute demonstration, which makes known both the existence and the cause of its object. But if what is unknown is the cause of the object, we shall only have a "demonstration *propter quid*."
>
> But the theory we are talking about belongs to neither of these modes of demonstration. For in this theory the principles are hidden from us, but they are in no way necessitated by the known effects. Astronomers are satisfied to assume such principles, although they do not know them.
>
> Furthermore, if you consider the effects that astronomers bear in mind when they advance their principles, you will find nothing in them from which it follows essentially and with necessity that this is the way things are. Having laid down unknown principles and having derived known conclusions from them, the astronomers have merely assumed the truth of the conversion.[7]

To propose mathematical hypotheses a priori, and then to derive conclusions from them that are faithful representations of the facts of observation, this, precisely, is—for the astronomer who is an adherent of Ptolemy—the essential task of anyone who constructs a theory. It would be quite absurd to think that experience, when it bears out the results of the deductions, transforms their premises into demonstrated truths. Nothing goes to show that altogether different premises might not have

6. Really to understand the meaning and importance of this argument, which figures so prominently in Duhem's *Essay*, the reader is advised to consult Heath's edition of Euclid's *Elements* (New York: Dover Publications, 1956). See index (in vol. 3), "analysis (and synthesis)."—TRANSLATOR.

7. Averroes *De Caelo* 2.35.

led to the same conclusions. Averroes is of course right to warn the astronomer of the error of overlooking this fact. But an astronomer who understands the true purpose of his science, as defined by men like Posidonius, Ptolemy, Proclus, and Simplicius, would not fall into this error, would not be trapped in the vicious circle of which the Commentator speaks; he would not require the hypotheses supporting his system to be *true*, that is, in conformity with the nature of things. For him it will be enough if the results of calculation agree with the results of observation—*if appearances are saved.*

But Averroes refuses to make do with this sort of astronomical theory. He requires that the science of the celestial motions take its principles from the teachings of physics, and from the only physics that is in his eyes true, namely, Aristotle's.

> The astronomer must, therefore, construct an astronomical system such that the celestial motions are yielded by it and that nothing that is from the standpoint of physics impossible is implied. . . . Ptolemy was unable to set astronomy on its true foundations. . . . The epicycle and the eccentric are impossible. We must, therefore, apply ourselves to a new investigation concerning that genuine astronomy whose foundations are principles of physics. . . . Actually, in our time astronomy is nonexistent; what we have is something that fits calculation but does not agree with what is.[8]

Averroes never found the leisure to undertake the task he considered necessary—the construction of an astronomical system that would not merely save the appearances but would rest on hypotheses which are in conformity with the nature of things, an astronomy based on principles drawn from the physics and metaphysics of Aristotle. "When I was young," he writes, "I hoped to be able to complete this investigation myself; now that I am old, I have abandoned that hope; but perhaps these words will induce someone else to undertake such a study."[9] Averroes' wish was fulfilled by his contemporary and fellow student al-Bitrogi.

Like Averroes a pupil of ibn-Tofail, like him a determined opponent of Ptolemy (whose work he seems to have known only in the form given it by Thabit ibn-Qurra), al-Bitrogi (Alpetragius) undertook to substitute a new system for the doctrines of the *Almagest.* Like Averroes, he claims that the principles upon which his theory of the planets rests can be proved by reasons drawn from physics; he goes so far as to call his

8. Averroes *Metaphysica,* lib. 12, summae secundae cap. 4, comm. 45.
9. Ibid.

treatise *The Theory of the Planets Proved by Physical Arguments* (*Planetarum theorica, physicis rationibus probata*).[10]

But, truth to tell, the metaphysics whose authority was invoked for the principles advanced by al-Bitrogi has only a very remote resemblance to the First Philosophy of Aristotle. Al-Bitrogi's metaphysics is directly derived from the *Liber de causis*, which the Arabs attributed to Aristotle and whose true origin was not known to the Christian Scholastics until the days of Thomas Aquinas, when the book was recognized as a medley of fragments borrowed from Proclus.

Even though the physics that supported it recalls the Academy far more than the Lyceum, al-Bitrogi's astronomy of homocentric spheres (in this respect akin to Aristotle's) was preferred to all others by the intransigeant Peripatetics of the later Middle Ages and the early Renaissance, who were more anxious to preserve the principles of "the Philosopher" and "the Commentator" than scrupulously to save the celestial phenomena.

Moreover, this system, beyond giving satisfaction to the faithful disciples of Averroes, who wished to base astronomy on hypotheses demonstrated by physics and conformable to the nature of things, appealed also to those whose imagination insisted on a theory that could be modeled by objects that an artisan might make of clay. And never before had this requirement been met by simpler expedients, since nine concentric spherical shells neatly fitted one inside the other represented the entire celestial machine.

Until the time of Copernicus, al-Bitrogi's essay and the attempts of his imitators would be competing with the Ptolemaic system for the favor of the Italian Averroists; and frequently the former would gain the upper hand.

It would appear then that the Arabs unanimously endorsed the axiom that astronomical hypotheses must conform to the nature of things. Some of them took this to mean that astronomical hypotheses must be deduced from a physics regarded as certain; others took it as referring to the condition that astronomical hypotheses be capable of representation by means of ingeniously sculpted and arranged rigid bodies. Not one of them seems to have risen to the doctrine that the Greek thinkers had enunciated; namely, that astronomical hypotheses are not judgments bearing on the

10. Al-Bitrogi, *Planetarum theorica, physicis rationibus probata, nuperrime latinis litteris mandata a Calo Calonymos Hebreo Neapolitano, ubi nititur salvare apparentias absque eccentricis et epicyclis.* Colophon: Venetiis, in aedibus Luce antonii Junte Florentini (January, 1801).

nature of things; that it is not necessary that they be deducible from the principles of physics, nor even that they be in harmony with these principles; that it is not necessary that they allow of representation by means of suitably arranged rigid bodies revolving on one another, because, as geometric fictions they have no function except that of saving the appearances.

Among works in Arabic, there is not one that contains the least glimmer of this Greek doctrine—with the important exception of the great twelfth-century treatise on philosophy and theology by the Jew Moses ben Maimon (Maimonides). There are several passages in this work—the *Guide of the Perplexed*[11]—where the learned rabbi explains his ideas about astronomical systems.

The idea that dominates in all of Maimonides' astronomical discussions —a new idea within Semitic Peripateticism, and one which, in this milieu, surprises by its sagaciously skeptical tendencies—is the one suggested by Ptolemy and developed by Proclus: The knowledge of heavenly things, in their essence and true nature, is beyond man's capacities; sublunary things alone are accessible to our feeble understanding:

I have promised you a chapter in which I shall expound to you the grave doubts that would affect whoever thinks that man has acquired knowledge as to the arrangement of the motions of the sphere and as to their being natural things going on according to the law of necessity, things whose order and arrangement are clear. I shall now explain this to you.[12]

In a discussion very similar to that of Averroes and al-Bitrogi, Maimonides then shows what, for one who is proficient in Peripatetic physics, is unacceptable about the epicycles and eccentrics assumed by astronomers.

Next he adds:

Consider now how great these difficulties are. If what Aristotle has stated with regard to natural science is true, there are no epicycles or eccentric circles and everything revolves round the center of the earth. But in that case how can the various motions of the stars come about? Is it in any way possible that motion should be on the one hand circular, uniform, and perfect, and that on the other hand the things that are observable should be observed in consequence of it, unless this be accounted for by making use of one of the two principles or of both of them? This consideration is all the stronger because of the fact that if one accepts everything stated by Ptolemy concerning the epicycle of the moon and its devi-

11. Moses Maimonides, *The Guide of the Perplexed*, ed. and trans. Shlomo Pines (Chicago: University of Chicago Press, 1963).

12. Ibid., pt. 2, chap. 23, p. 322.

ation toward a point outside the center of the world and also outside the center of the eccentric circle, it will be found that what is calculated on the hypothesis of the two principles is not at fault by even a minute. . . . Furthermore, how can one conceive the retrogradation of a star, together with its other motions, without assuming the existence of an epicycle? On the other hand, how can one imagine a rolling motion in the heavens or a motion around a center that is not immobile? This is the true perplexity.[13]

How is a man to extricate himself from this perplexity? In the way suggested by Posidonius, Geminus, Ptolemy, and Simplicius. Maimonides adopts the doctrines of these Greek thinkers, and the terms they had used to give expression to their thought are almost identical with his.

Consider, for example, the following passage, which mentions only Ptolemy by name, but sounds as if Simplicius himself were speaking:

Know with regard to the astronomical matters mentioned that if an exclusively mathematical-minded man reads and understands them, he will think that they form a cogent demonstration that the form and number of the spheres is as stated. Now things are not like this, and this is not what is sought in the science of astronomy. Some of these matters are indeed founded on the demonstration that they are that way. Thus it has been demonstrated that the path of the sun is inclined against the equator. About this there is no doubt. But there has been no demonstration whether the sun has an eccentric sphere or an epicycle. Now the master of astronomy does not mind this, for the object of that science is to suppose as a hypothesis an arrangement that renders it possible for the motion of the star to be uniform and circular with no acceleration or deceleration or change in it and to have the inferences necessarily following from the assumption of that motion agree with what is observed. At the same time the astronomer seeks, as much as possible, to diminish motions and the number of the spheres. For if we assume, for instance, that we suppose as a hypothesis an arrangement by means of which the observations regarding the motions of one particular star can be accounted for through the assumption of three spheres, and another arrangement by means of which the same observations are accounted for through the assumption of four spheres, it is preferable for us to rely on the arrangement postulating the lesser number of motions. For this reason we have chosen in the case of the sun the hypothesis of eccentricity, as Ptolemy mentions, rather than that of an epicycle.[14]

Why is the astronomer powerless to transform his hypotheses into demonstrated truths? The reason is the limited character of human sci-

13. Ibid., pt. 2, chap. 24, pp. 325–26.
14. Ibid., pt. 2, chap. 11, pp. 273–74.

ence, which cannot attain to knowledge of heavenly things. Ptolemy had insinuated this explanation; Proclus stated it more fully; and Maimonides reiterates it:

I shall repeat here what I have said before. All that Aristotle states about that which is beneath the sphere of the moon is in accordance with reasoning; these are things that have a known cause, that follow one upon the other, and concerning which it is clear and manifest at what points wisdom and natural providence are effective. However, regarding all that is in the heavens, man grasps nothing but a small measure of what is mathematical; and you know what is in it. I shall accordingly say in the manner of poetical preciousness: *The heavens are the heavens of the Lord, but the earth hath He given to the sons of man* [Ps. 114:16]. I mean thereby that the deity alone fully knows the true reality, the nature, the substance, the form, the motions, and the causes of the heavens. But He has enabled man to have knowledge of what is beneath the heavens, for that is his world and his dwelling-place in which he has been placed and of which he himself is a part. This is the truth. For it is impossible for us to accede to the points starting from which conclusions may be drawn about the heavens; for the latter are too far away from us and too high in place and in rank. . . . And to fatigue the minds with notions that cannot be grasped by them and for the grasp of which they have no instrument, is a defect in one's inborn disposition or some sort of temptation.[15]

The effort to establish a sublunary physics that teaches us about the real properties of the elements and their compounds is reasonable; but it is madness to attempt to construct a celestial physics that would claim, by means of its principles, to know the quintessence. Such is Maimonides' conclusion.

15. Ibid., pt. 2, chap. 24, pp. 326–27.

3
Medieval Christian Scholasticism

Which astronomical doctrine had one best adopt? Is Ptolemy's system to be used, or al-Bitrogi's theory to be relied on? The geometric constructions of the *Almagest* are marvelously suited to saving the phenomena. Using these constructions, the calculators have been able to set up tables which predict the tiniest detail of the celestial motions, and the discrepancies between the entries on these tables and the facts of observation have been imperceptibly small. But the hypotheses on which these constructions are based were not set up in conformity with Peripatetic physics; what is more, this physics yields arguments that tend to overturn the Ptolemaic hypotheses. Al-Bitrogi's theory, on the other hand, does duly respect the physics of Aristotle (i.e., what it takes to be such), but its deductions leave off long before yielding results that can be compared with observation. There is no way of telling whether this doctrine is capable of saving the phenomena: its deductions are simply not pushed far enough to allow for the construction of astronomical tables and almanacs.

Between these two astronomical systems the Christian Scholasticism of the thirteenth century hung suspended—urged in one direction by its lively curiosity, which created a desire for a natural science in conformity with the lessons of experience; drawn to the other by its respect for the metaphysics of "the Philosopher." Among the Scholastic doctors there were some few who recognized that the solution of the dilemma turns on the question of the value properly assigned to astronomical hypotheses.

Bernard of Verdun, in his *Tractatus super totam astrologiam*,[1] gives a very full account of the debate between the two systems, then decides in favor of Ptolemy's. For him the hypotheses supporting this system are true; their truth is demonstrated by the fact that the propositions they entail have for ever so long been found to agree with the observed motions; in his opinion the hypotheses should be considered truths of fact: their certainty is an immediate consequence of sense experience and eludes demonstration because it is prior to all demonstration and rules it:

The first way [al-Bitrogi's theory] is impossible. It is insufficient for saving the phenomena previously enumerated—phenomena which every reasonable man is bound to concede. It remains therefore that the second way, the one which consists in assuming an eccentric, an epicycle, and numerous orbits . . . is necessary. On this theory all the disadvantages we have just been talking about are avoided and the appearances listed in the preceding chapter are saved. By adopting it as our starting point we are able to determine and predict everything that can possibly be known about the celestial motions as well as the distances and magnitudes of the celestial bodies. And up to our time these predictions have proved exact; which could not have happened if this principle had been false; for, in every department, a small error in the beginning becomes a big one in the end.

Everything that appears in the heavens agrees with this theory and contradicts the other. And just as it is necessary to defend the truths of observation previously enumerated, so it is necessary to concede the correctness of the present theory, and this by the same necessity that forces us to admit the celestial movements in all nature. On the strengh of a few sophistical arguments, to deny what is more certain than all argument is absurd; it is a folly similar to that of those ancients who, because of a few sophisms, denied movement and every kind of change and the plurality of beings—things the falsity and contradiction of which are manifest to our senses. These things cannot be demonstrated, just as one cannot be demonstrated that fire is hot or that everything that exists involves substance and accident. Sensation assures us that this is the way things are. Thus the Philosopher states that we know these things with greater certitude than any argument could provide; and he adds that it would not be right to look for arguments for them because all argument on our part presupposes sense perception.[2]

No matter how numerous and exact the confirmations that are brought to a theory by experience, the hypotheses supporting the theory never at-

1. Paris, Bibliothèque nationale, fonds latin, ms. nos. 7333, 7334 (Bernard of Verdun "Tractatus optimus super totam astrologiam").

2. Ibid., dist. 3, cap. 4.

tain to the certainty of commonsense truths. It would be a serious mistake to think that they do, and Bernard of Verdun, who did think so, was to that extent quite naïve. To defend such a position in our day, after history has witnessed the collapse of so many theories long accepted without dispute, one would have to be even more naïve. Yet how many of our contemporaries, the very ones who think of themselves as tough-minded, accord scientific theories the same unquestioning confidence as did the humble Franciscan friar of the thirteenth century.

It is absurd to believe that experimental control can transform the hypotheses upon which a theory rests into truths of immediate perception. It is still more absurd to fix upon a metaphysical system to the point of upholding its conclusions in spite of refutation by experience. Yet this is the extreme to which, it seems, Roger Bacon let himself be carried.

We know that Roger Bacon had included an account of the motions of the heavenly bodies in a part (now lost) of his *Opus minus.* Certain considerations (too far removed from our present subject matter to find a place in this essay) lead us to identify this account with a fragment belonging to a miscellany Bacon assembled from various parts of his previous works, which he called *Communia naturalia.*

The first of the three chapters which make up this fragment[3] is concerned with the two systems of astronomy, Ptolemy's and al-Bitrogi's. This chapter, which is interlarded with typical passages from Bernard of Verdun's *Tractatus super totam astrologiam*, has the appearance of a polemic against the latter work:[4]

Those people who mean to destroy the epicycles and eccentrics say that it is better to save the order of nature and to contradict sensation, which is so often found to be in error, especially in cases involving great distances. In their opinion it is better to leave unsolved a sophism that is hard to answer than to assume knowingly what is contrary to nature.

Pursuing his account, Bacon returns to this idea, and now he appears to adopt it as his own:

Mathematical physicists, those who follow the ways of nature, try, no doubt, as do pure mathematicians, who do not know physics, to save the

3. Paris, Bibliothèque Mazarine, ms. no. 3576, fol. 130 ("Incipit liber primus communium naturalium Fratris Rogeri Bacon. . . . Incipit secundus liber communium naturalium, qui est de celestibus, aut de caelo et mundo. . . . Incipit quinta pars secundi libri naturalium. . . ," chap. 17).

4. More recent studies, supervening upon the completion of the present essay, have convinced us that Roger Bacon's written works preceded and inspired Bernard of Verdun's *Treatise.* Nevertheless, what is said above about the opposition of the two Franciscans' opinions holds.

appearances. But they try at the same time to save the order and the principles of physics, whereas pure mathematicians destroy them. It seems, therefore, that we had better imitate the physicists in our assumptions even though this means that we shall be found wanting in the solution of certain sophisms drawn less from reason than from sensation.

The man who wrote these lines is the same who is so often represented as the redoubtable adversary of the deductive cosmology of the Peripatetics, as the precursor of the experimental method!

Despite his predilection for an astronomy founded on the principles of physics (i.e. the astronomy of al-Bitrogi)—a predilection that refuses to yield even when the facts are against it—Bacon had to acknowledge that "there are no instruments, no canons, no astronomical tables constructed with a view to submitting the physicitsts' hypotheses to the test of fact." He is forced to admit that the object of any astronomical theory is to furnish calculations that conform to observation:

Here is something that must be known and deserves attention: Although pure mathematicians and those who know physics propose different methods for saving what appears in the celestial bodies, nevertheless they are all headed toward one and the same goal, albeit along different roads: the goal is to know the positions of the fixed stars and the planets in relation to the Zodiac; so while they disagree about what road is to be followed, they are together in acknowledging that it must terminate at this goal and limit.

It would surely have been very odd if Bacon had stuck to the position he momentarily adopted in the previously quoted passage from the *Opus minus*, if he had persisted in placing al-Bitrogi's cosmology beyond the test of experience. Actually, he was not long in abandoning that foolish position.

We have recently uncovered a hitherto unknown, and very important, portion of the *Opus tertium*, Bacon's last work, which he dedicated to Pope Clement IV. In the manuscript preserving it for us,[5] this section is entitled: "The Third Book of al-Bitrogi, in which he discusses perspective: On the comparison of science with wisdom: On the movements of celestial bodies according to Ptolemy. On al-Bitrogi's opinion contrary to Ptolemy's and others' opinions. On the science of experiments concerning nature: On the comportment of scientists. On alchemy."[6] As the title

5. Paris, Bibliothèque nationale, fonds latin, ms. no. 10264, fols. 186 (recto)–220 (recto).

6. "Liber tertius Alpetragii. In quo tractat de perspectiva: De comparatione scientie ad sapientiam: De motibus corporum celestium secundum Ptolomeum. De opinione Alpetragii contra opinionem. Ptolomei et aliorum. De scientia experimentorum naturalium. De scientium morali. De Alkimia."

suggests, this fragment of the *Opus tertium* contains a long discussion of the astronomical systems of Ptolemy and al-Bitrogi. Bacon introduced an only slightly altered version of this discussion into his *Communia naturalium*,[7] placing it immediately before the discussion drawn from the *Opus minus*, untroubled by the flagrant contradiction introduced into his presentation by this inversion of the chronological sequence of the two fragments.

The later fragment often lets us catch glimpses of Bacon's desire to decide in favor of the doctrine of al-Bitrogi; nevertheless he ends by recognizing[8] that the doctrine is incompatible with a certain number of facts, facts which in the *Opus minus* he had written off as "sophisms"; they are the very facts which Bernard of Verdun had entered on the list of "essential appearances," to be saved by *any* astronomical theory.

The vacillations of Roger Bacon show, no less than the incautious certainties of Bernard of Verdun, how little these two Franciscans understood the true nature of astronomical theories. The wisdom of, say, a Simplicius never reached them, it seems.

Their confrère—later to be canonized—Bonaventura does seem to have glimpsed a sort of reflection of that wisdom; and he uses it against those who, made confident by the confirmations of experience, claim to transform the system of Ptolemy into demonstrated truth:

> To the sense it seems that the supposition of the mathematicians is the most correct, since the deductions and judgments which are based on that supposition do not lead to a single erroneous result concerning the motions of the heavenly bodies. All the same, according to reality it is not necessary that this, the mathematician's, position be more true (*secundum rem tamen non oportet esse verius*), for the false is frequently a means of discovering the truth; it would seem that the philosopher of nature makes use of a more reasonable method and supposition.[9]

Caught between the system of Ptolemy, which saves the appearances by rejecting the principles of Peripatetic physics, and the system of homocentric spheres, which relies on these principles but does not square with the facts, Bonaventura does not know which side to choose; at this point he recalls the teaching of the Greek thinkers. They taught that observa-

7. Paris, Bibliothèque Mazarine, ms. no. 3576, fols. 120–130 (Roger Bacon "Communium naturalium" 2.5–2.6).

8. Paris, Bibliothèque nationale, fonds latin, ms. no. 10264; Paris, Bibliothèque Mazarine, ms. no. 3576, fol. 129.

9. Bonaventura, *In secundum librum Sententiarum disputata* 14.2.2, "Utrum luminaria moveantur in orbibus suis motibus propriis."

tion's agreement with the logical consequents of a theory does not certify the truth of the hypotheses on which the theory is based; the appearances migh conceivably be saved by means of other hypotheses. Bonaventura therefore hopes for the invention of some new system that would save both the principles of the physicist and the observations of the astronomer.

This hope for some future more adequate system is perhaps hardly articulate in the work of the Seraphic Doctor, but in the writings of the Angelic Doctor it finds firm expression.

In his lessons on the *De caelo et mundo*, Thomas Aquinas sketches the philosophical foundations of an astronomical theory that admits only uniform rotations around the center of the universe. This theory saves all the principles of Peripatetic metaphysics. Does it also agree with astronomical observation? Aquinas is well aware that it does not.[10] Even Eudoxus, Calippus, and Aristotle in their day had been forced to complicate the system of homocentric spheres outrageously in order to let it represent the various accidents of the courses of the planets, and several of these complications find no justification whatever in the Aristotelian philosophy. That the eccentrics and epicycles envisaged by Hipparchus and Ptolemy are not grounded in Aristotle's philosophy is all the more obvious.

What credence do the hypotheses that underlie the various astronomical systems deserve? Averroes had already insisted that the reasoning whereby the geometers attempt to justify hypotheses does not amount to anything like a demonstration. Averroes' critique seems to have inspired Aquinas to the following reflection:

> The assumptions of the astronomers are not necessarily true. Although these hypotheses appear to save the phenomena (*salvare apparentias*), one ought not affirm that they are true, for one might conceivably be able to explain the apparent motions of the stars in some other way of which men have not as yet thought.

This was an idea that Aquinas had, in fact, expressed even earlier, albeit somewhat more concisely, in the course of explaining Aristotle's fundamental axiom that all simple circular motion is around the world's center.[11]

> For a wheel which moves around its own center does not move with a purely circular movement; its movement is complicated by a rising and a descending motion.

10. Thomas Aquinas, *Expositio super librum de Caelo et Mundo* 2.17.
11. Ibid., 1.3.

But it seems, according to this suggestion, that the heavenly bodies are not all moved by a circular movement. For according to Ptolemy planetary movements are effected through epicycles and eccentrics, and these movements do not take place around the center of the universe, which is the earth's center; they take place around certain other centers.

It must be noted in this connection that Aristotle does not admit that this is the way things are. He assumed, with the astronomers of his time, that all heavenly movements are described around the center of the earth. Later, Hipparchus and Ptolemy thought up the movements of eccentrics and epicycles in order to save what is manifest to sense in the heavenly bodies. This, then, is not something proved—it is merely an assumption (*unde hoc non est demonstratum, sed suppositio quaedam*). But if such an assumption were true, the heavenly bodies would continue to move around the center of the universe by a diurnal movement, which is the movement of the supreme sphere, the one which carries all the heavens along in its revolution.

The hypotheses which support an astronomical system are not transformed into demonstrated truths by the mere fact that their conclusions agree with observation. Thomas Aquinas makes this assertion following Averroes, although in a less severe style. And this precept of logic must have seemed quite important to him, since he repeats it again in another place:[12]

We can account for a thing in two different ways. The first way consists in establishing by a sufficient demonstration that a principle from which the thing follows is correct. Thus, in physics we supply a reason which is sufficient to prove the uniformity of the motion of the heavens. The second way of accounting for a thing consists, not in demonstrating its principles by a sufficient proof, but in showing which effects agree with a principle laid down beforehand. Thus, in astronomy we account for eccentrics and epicycles by the fact that we can save the sensible appearances of the heavenly motions by this hypothesis. But this is not a really probative reason, since the apparent movements can, perhaps, be saved by means of some other hypothesis.

In these various texts Aquinas is adopting ideas that we have heard expressed before by Simplicius, whose language he very nearly borrows. We here find a clear indication of the Greek commentator's influence on the Scholastic commentator. And Aquinas' lectures on the *De caelo et mundo* show that we are not dealing with a mere coincidence—he there on several occasions cites Simplicius' commentary on the same book.[13]

12. Thomas Aquinas *Summa Theologica* 1.32.1-2.
13. See especially, bk. 1, lect. 6; and bk. 2, lect. 4.

The principles that Thomas Aquinas, following Simplicius, laid down enabled astronomers to use Ptolemy's hypotheses without scruple in their study of the apparent movements of the planets, despite the fact that their metaphysical opinions might force them to reject these hypotheses. Thus John of Jandun, a great admirer both of Aristotle and of Averroes, nevertheless adopts, along with all the astronomers of his time, the only astronomical theory which furnished observers and calculators with canons and almanacs. "With Ptolemy and all the modern astronomers" he declares that it is necessary to assume the existence of eccentrics and epicycles:[14]

For we must admit, with respect to the celestial bodies, the hypotheses that enable us to save the phenomena (*salvare apparentias*) if, without recourse to these hypotheses, we cannot save and account for these phenomena, which have for so long been observed and confirmed without risk of error.

But does the fact that "these orbits precisely determine the places and movements of the planets, that they are exactly right for computational purposes and for the construction of astronomical tables" go to show that they have a real and essential existence, *in esse et secundum rem?* It hardly matters to the astronomer.

For him it suffices to know the following: If the epicycles and eccentrics did exist, the celestial motions and the other phenomena would occur just as they do now. The truth of the conditional is what matters, whether or not such orbits really exist among the heavenly bodies. The assumption of such eccentrics and epicycles is sufficient for the astronomer *qua* astronomer because as such he need not trouble himself with the reason why (*unde*). Provided he has the means of correctly determining the places and motions of the planets, he does not inquire whether or not this means that there really are such orbits as he assumes up in the sky: *that* investigation concerns the physicist. For a consequence can be true even when its antecedent is false.[15]

14. John of Jandun *Acutissimae quaestiones in duodecim libros Metaphysicae ad Aristotelis et magni Commentatoris intentionem ab eodem exactissime disputatae* 12.20.

15. For a moderately detailed account of the scholastic "theory of consequence" here in evidence, i.e. the scholastic analysis of the logical import of the 'if . . . then' connection in "hypotheticals" like the ones that figure in astronomical theory, see E. A. Moody, *Truth and Consequence in Medieval Logic* (Amsterdam: North-Holland Publishing Company, 1953). It was, perhaps, under the influence of Duhem's thesis that "modern" mechanics, i.e. the "new science" of Galileo, really goes back to the fourteenth century Scholastics at the University of Paris, that "modern" logicians began to look for the ancestry of their own "formal logic" in the logical textbooks used at Paris and Oxford.—TRANSLATOR.

These lines were written at the University of Paris some time around the year 1330!

The end of the Middle Ages slips by without that university's providing us, through its teaching, with any new documents concerning the value of astronomical hypotheses. Astronomy was going through one of those periods of quiet possession when no need is felt to discuss the principles that underlie theory, when all are directing their effort to working out the applications of theory. In the fourteenth century, at Paris, the system of Ptolemy was accepted without argument.

The Italian schools of the same period yield little that is of interest. At the time the study of astronomy was less advanced there than at Paris. Interest ran particularly to astrology, and the nature and value of the hypotheses it too employs was hardly discussed.

There is, however, one exception—Peter of Padua (Peter of Abano).

Some time after the year 1303 Peter of Padua completed his celebrated *Conciliator differentiarum philosophorum et medicorum*, a work that became extremely popular and won him the nickname of *Petrus Conciliator.* He had formed the project of writing an analogous work on astronomy, to be entitled *Lucidator astronomiae.* Whether the project was ever carried out we do not know. The Bibliothèque nationale has the manuscript of the *Proemium* and of the first chapters[16] or, as Peter of Padua refers to them, "the first distinctions." This fragment was written in 1310. Unfortunately the copyist, a certain Peter Collensis, was as clumsy a scribe as he was an illiterate Latinist.

Peter of Padua was a compiler. We should not expect him to make his logical position perfectly firm and clear. Nevertheless, when he discusses the hypotheses bearing on eccentrics and epicycles, he is obviously guided by the doctrines of a single philosophy, namely Ptolemy's.[17]

He recalls that:

In Aristotle's and Ptolemy's opinion, nature and art always strive to arrive at their ends by the shortest means, and it is a mistake to effect through many means what can be achieved through fewer, as is shown in the first book of the *Physics.* According to Ptolemy, the hypotheses of eccentrics and epicycles conform to this principle, since every action of the celestial machine is completely reproduced by using only eighteen motions.[18]

16. Paris, Bibliothèque nationale, fonds latin, ms. no. 2598, fols. 99 (recto)–125 (verso).

17. Peter of Padua *Lucidator Astronomiae,* diff. 4 a, "An sit ponere eccentricos et epicyclos?" fol. 112, col. c–fol. 116, col. c.

18. Ibid., fol. 112, col. c.

Having explained the various systems proposed by astronomers at length, he adds:

So we have briefly shown that none of the preceding opinions can completely save what appears to the astronomer, but that some systems involve results more absurd than those which follow from the others.[19]

We should yield preference to the system of Ptolemy and his disciples, who assumed eccentrics and epicycles, because they sufficiently account for the appearances and do so by the smallest number of motions.

Peter of Padua then invokes the authority of Simplicius, after which he continues as follows:

What confirms me in this assumption is that it [Ptolemy's system] uses the least number of instruments to represent the heavenly movements: For I think that we should not compound this motion from many elements when we can construct it more directly and more quickly. The arts show the justice of this consideration. Besides, this assumption—as can be seen with the help of instruments—is better at saving the appearances. And finally, it is more successful than the others in discovering the periods of the revolutions of the spheres and the planets through computation.[20]

The quoted passages from the writings of Peter of Padua summarize the philosophy of science of the medieval Christian astronomers. This philosophy is further summarized in two principles:

The hypotheses of astronomy should be as simple as possible.

They should save the phenomena as exactly as possible.

19. Ibid., fol. 115, col. a.
20. Ibid., col. b.

4
The Renaissance before Copernicus

In the fourteenth century, the University of Paris took to "swarming." Of the many learned Masters of the English Nation (who hailed for the most part from German-speaking countries) some would now occasionally leave the banks of the Seine to found new universities in German lands. The latter were like colonies of their alma mater; the German-speaking universities tended to remain exposed to currents of thought emanating from Paris.

Thus it was that some time around 1380 Heinrich Heimbuch of Hesse, a very learned Master of Arts and Bachelor of Theology, left the schools on the Rue du Fouarre and the lecterns at the Sorbornne to become the "Planter" of the University of Vienna (*Plantator gymnasii Viennensis,* as he has often been called). Astronomer as well as theologian, he oriented the new university along the lines that had been impressed upon him by his teachers. Accepting the principles of the Ptolemaic system unquestioningly, the school of Vienna directed all its efforts to working out the details of astronomical theory; new procedures for calculation were invented and old ones perfected, astronomical tables and almanacs prepared, instruments constructed, methods of observation devised. Its most illustrious teachers, men like Georg Purbach or Regiomontanus (Johann Müller of Königsberg) were perfect examples of the type of investigator who excels in all the technical details of a science but who never dreams of examining the nature and value of the hypotheses which support his science.

While the Viennese astronomers were classifying the postulates of the

Ptolemaic system as truths established once and for all, the Averroists of the School of Padua, fanatical admirers of the teachings of "the Commentator," were conducting frenzied attacks upon these doctrines.

Like their master, the Italian Averroists refused astronomy the right to use hypotheses that do not conform to the nature of things—the physics of "the Philosopher" and "the Commentator." Like Averroes, they declared the Ptolemaic system unacceptable on this head. And like al-Bitrogi, those among them who considered themselves astronomers tried to substitute for the theory of the *Almagest* one founded exclusively on the use of homocentric spheres.

Nicholas of Cusa, who had studied at Padua, made a preliminary attempt in this direction. But he had the sense to keep it a secret. Alessandro Achillini, the famous rival of Pomponazzi, saw no reason to emulate Cusa's caution. In 1494 he had the *Quatuor libri de orbibus* printed in Bologna. A new, corrected edition of the book was published in 1498. And it was reprinted in the collected works of Achillini published in Venice in 1508, 1545, 1561, and 1568.

The *Four Books on the Spheres* painstakingly and in pedantic detail develop Averroes' doctrine concerning the material of the heavens, their form, and the intelligences by which they are moved.

It is in the first book of this work[1] that the celebrated Averroist professor of Bologna and Padua undertakes to destroy the Ptolemaic system and presents a sketch of the theory he would put in its place.

At the very outset, Achillini submits the following proposition:

> The motions that Ptolemy assumes are founded upon two hypotheses, the eccentric and the epicyclic, which do not agree with physics. Both these hypotheses are false.[2]

Against the hypotheses in question he urges all the old arguments of Averroes. He suggests the basis of an astronomical doctrine that is in his opinion in conformity with the principles of sound physics; this astronomy, it turns out, hardly differs from al-Bitrogi's. Repeating a remark of Averroes word for word, he comes to the conclusion that:

> A real astronomy is nonexistent. [What passes for astronomy] is merely something suitable for computing the entries in astronomical almanacs.[3]

1. Alessandro Achillini, *Liber primus de orbibus*, dubium tertium, "An eccentrici sunt ponendi."
2. Achillini, *Opera omnia* (Venice: apud Hieronymum Scotum, 1545), fol. 29, col. b.
3. Ibid., fol. 31, col. b.

But, writes Achillini, it will be objected that:

One must needs concede hypotheses which for very long and without any error have stood the test of observation and without which it is impossible to save the phenomena. And the eccentrics and epicycles fit this description.[4]

To which Achillini replies:

The minor of this syllogism must be denied, since we propose to account for phenomena by other causes. Moreover, astronomers have not established the existence of eccentrics and epicycles by any sort of demonstration. . . . Plainly, they have not proved their existence a priori; but neither have they proved it a posteriori; for the effects that are manifest to us may stem from other causes. . . . Ptolemy commits an error in physics when he uses fictional bodies as causes by which to account for phenomena.

Agostino Nifo likewise attempted to write an astronomy from which the hypotheses of Ptolemy were ousted.[5] The geometric constructs of those who accept these hypotheses he writes off as "old wives' tales." Against them and against their doctrines, he too, like Achillini, advances the various arguments of Averroes:

You must understand that a good demonstration proves that the cause necessitates the effect, and conversely. Now it is true enough that, the eccentrics and epicycles being conceded, the observed phenomena follow and that they can, therefore, be saved in this way. But the converse does not hold. Starting with the appearances, the existence of eccentrics and epicycles does not follow with necessity; only provisionally, until such time as a better cause be discovered, one which both necessitates the phenomena and is necessitated by them, are the eccentrics and epicycles established. Those therefore err who, starting out from a proposition whose truth may be the outcome of various causes, decide definitively in favor of one of these causes. The appearances can be saved by the sort of hypotheses we have been talking about, but they may also be saved by other suppositions not yet invented.

. . . There are three species of demonstration: *demonstration by signs*—from an effect that is known to us, we infer the cause of that effect; *demonstration by the cause only*—from the cause, which we have been able to discover through its effect, we infer the effect; *demonstration by the cause and essence jointly*, also called demonstration in the strict sense or *de*

4. Ibid., fol. 35, coll. a, b.

5. *Aristotelis Stagiritae De Caelo et Mundo libri quatuor e Graeco in Latinum ab Augustino Nipho philosopho Suessano conversi, et ab eodem etiam praeclara, neque non longe omnibus aliis in hac scientia resolutiore aucti expositione. . . .* (Venice: apud Hieronymum Scotum, 1549), bk. 2, fol. 82, cols. c, d.

natura—which starts from a cause that is known to us; such is geometric demonstration.

Now the proof of the existence of eccentrics and epicycles belongs to none of these three kinds of demonstration:

From the apparent motions one can easily move to the eccentrics and epicycles. But it is impossible to go in the reverse direction, since one would then be moving from the unknown to the known, the appearances being what is known to us, the eccentrics and epicycles unknown.

The quoted passages are not a mere repetition of Averroes' teachings. They also bear the imprint, and recognizably so, of the thought of Thomas Aquinas, some of whose statements are reproduced word for word in Nifo's exposition.

Nifo's critique successfully proves that a theory's harmony with observation cannot transform the hypotheses upon which it rests into demonstrated truths. *Demonstration* would require that one establish in addition that no other set of hypotheses is capable of saving the phenomena. But the Renaissance Averroists of Padua failed to draw the sensible conclusion which John of Jandun, in the fourteenth century, had transmitted to them from Paris. They did not grant astronomy the right to make use of purely fictive but convenient hypotheses. They did not want it to limit its ambitions to the construction of a theory on which the entries in astronomical tables and almanacs might be based. Worthy successors of the Commentator and his codisciple al-Bitrogi, they insisted on constructing astronomy on the basis of principles demonstrated by physics and would severely condemn whosoever might claim to be proceeding in a different manner.

Listen, for instance, to Fracastoro, as he presents his book *Homocentrics* to Pope Paul III (1535):

You are well aware that those who make profession of astronomy have always found it extremely difficult to account for the appearances presented by the planets. For there are two ways of accounting for them: the one proceeds by means of those spheres called homocentric, the other by means of so-called eccentric spheres. Each of these methods has its dangers, each its stumbling blocks. Those who employ homocentric spheres never manage to arrive at an explanation of phenomena. Those who use eccentric spheres do, it is true, seem to explain the phenomena more adequately, but their conception of these divine bodies is erroneous, one might almost say impious, for they ascribe positions and shapes to them that are not fit for the heavens. We know that, among the ancients, Eudoxus and Calippus were misled many times by these difficulties. Hipparchus was among the first who chose rather to admit eccentric spheres than to be found wanting by the phenomena. Ptolemy followed him, and soon practically all astronomers were won over by Ptolemy. But against these astronomers, or at least, against the hypothesis of eccentrics, the whole of philosophy has raised continuing protest. What am I saying? Philosophy? Nature and the celestial spheres themselves protest unceasingly. Until now, no philosopher has ever been found who would allow that these monstrous spheres exist among the divine and perfect bodies.[6]

Fracastoro is not satisfied merely to avoid these absurd hypotheses; nor will it do for him to set up a theory that will serve the purposes of calculation; he claims to have hit on the very causes of the celestial movements:

In our *Homocentrics* not merely that utility will be found which results from any astronomical theory; other things that are greatly to be desired will be found in it as well. These things seem, in the first place, to contribute much to truth, the object we ought to desire the most; they contribute to the discovery of the proper causes of the celestial movements and, finally, to the [discovery of] the very qualities of these movements.[7]

One year after the publication of Fracastoro's *Homocentrics*, Gianbattista Amico issued his *opusculum* on the same subject. In the first chapter of this little book he tells us:

Among the ancients there were some who attempted to unite astronomy with natural philosophy; others, on the contrary, strove to separate the two sciences. Eudoxus, Calippus, and Aristotle sought to reduce all the varied and non-uniform movements which the heavenly bodies display to homocentric spheres such as nature sanctions. Ptolemy, on the other hand, and those who followed his method, wanted, in spite of nature, to reduce these same movements to eccentrics and epicycles.[8]

Astronomers attribute the phenomena we perceive when observing the celestial bodies to eccentrics and to those little spheres that are called epicycles. But they have done a bad job of reducing all these effects to such causes. Besides, it is hardly surprising that they should have fallen into error in making this reduction. As Aristotle says in the first book of the *Posterior Analytics*, any solution is difficult when those who advance it use false principles. If, then, nature knows neither epicycles nor eccentrics, which is the position taken, and quite properly, by Averroes, . . . it behooves us to reject these spheres. And we do so all the more willingly seeing that astronomers ascribe to the epicycles and eccentrics certain motions that they call "inclinations," "reflections," "deviations"—move-

6. Hieronymus Fracastor *Homocentricorum, sive de stellis, liber unus* (Venice, 1535).

7. Ibid., chap. 1.

8. Gianbattista Amico *De motibus corporum coelestium juxta principia peripatetica sine eccentricis et epicyclis* (Venice, 1536), chap. 1.

ments that cannot, at least in my opinion, belong to the quintessence at all.[9]

In every age we find people who believe that they are able to penetrate to the very nature of bodies and who think that they can discover such truths concerning this nature that physics becomes "deducible" as from its first principles. It has almost always been impossible to compel such physicist-philosophers to push their deductions through to the end, to develop their theory to the point where its consequences can be subjected to the test of experience.

The Averroists loudly proclaimed that they were in possession of those physical truths from which any admissible astronomy proceeds.

Like al-Bitrogi, they sketched out the plan of the theory they proposed to construct on these foundations. But, again like al-Bitrogi, they were careful never actually to erect the projected edifice. They did not particularize their system to the point where it might be reduced to astronomical tables and the information contained by these tables compared with the statements of observers. Thus Alessandro Achillini writes:

We have no intention of explaining what, according to our assumption, the proper causes of all the variations of the celestial movements are. This is a task which we must leave to astronomers. Led by the hand, as it were, of what we have said, they will, I am confident, know how to investigate and work out everything to the point of furnishing this complement to our theory.[10]

This is what Fracastoro says on the same subject:

In accounting for the movements of the planets, we have passed over calculations of extreme intricacy, and no one will find this surprising. For we believe that these calculations do not really pertain to our work. We grant that such minute estimates should be expected of astronomical tables, but the tables now used can easily be accommodated to our homocentrics.[11]

And Gianbattista Amico declares:

In this work one will, perhaps, find nothing completed, but I believe that I have done enough if I have been able to arouse more eminent minds to the desire to make this explanation clearer.[12]

9. Ibid., chap. 7.

10. Achillini *De orbibus liber primus* (end).

11. Fracastor *Homocentricorum, sive de stellis, liber unus* (end of last chapter).

12. Amico to Cardinal Nicolaus Rodulphus, "De motibus corporum coelestium."

The Averroists refused to concede that astronomy has achieved its purpose when it has succeeded in saving the appearances. Still, they never dared deny that it has to agree with the phenomena. But they were never in a position to check whether or not their own theory complied with this condition.

If the Averroists were victims of the illusion that one can deduce an astronomical theory from a metaphysical doctrine, the partisans of the Ptolemaic system sometimes let themselves be seduced by another illusion: They thought that the exact determination of appearances can bestow certainty on the assumptions that are designed to account for the observed facts. By opposite paths, Averroists and Ptolemaists both ended up in the same error: that of ascribing independent reality to the hypotheses upon which astronomical theory rests.

Francesco Capuano of Manfredonia (or of Maria Siponto), professor of astronomy at the University of Padua, who changed his given name to Giovanni Battista when he left the world to enter the order of the Canons Regular of the Lateran, fell victim to the second illusion.

In 1495, he had a *Commentary* on the *Theory of the Planets* of Georg Purbach printed in Venice.[13] This commentary was later issued in a large number of editions.

Capuano devotes several pages of the book to the rebuttal of Averroist objections to the epicycles and eccentrics. The objections answered are not only those of Averroes, but also those which had been addressed to Capuano personally by "an ingenious imitator of Averroes belonging to our own time and country" (quidam subtilis hujus aetatis, et noster conterraneus Averrois imitator)—and by these words Capuano is surely referring to his colleague Achillini.

For the commentator on Purbach, establishing the tenability of the Ptolemaic hypotheses is not sufficient; he wants them to be true; and he proposes to prove them; not, indeed, a priori, but at least a posteriori.

In his introduction he announces that he will "demonstrate a priori all that is susceptible of a priori and mathematical proof and that, as for the principles, such as the spheres and their movements, of which no such demonstration can be given," he has decided "to let them be known a

13. *Theorice nove planetarum Georgii Purbachii astronomi celebratissimi, ac in eas eximii artium et medicine doctoris Domini Francisci Capuani de Manfredonia in studio Patavino astronomiam publice legentis sublimis expositio et luculentissimum scriptum* (Venice: per Simonem Bevilaquam Papiensem. August 10, 1495).

posteriori and by means of the appearances." A little later he expands on this thought:

Here, as in the *Almagest*, the roads leading to science are the two kinds of demonstration—demonstration by signs and demonstration in the strict sense. Now the principles of astronomy are inferred, a posteriori and from sense: having noted and observed the motion of a planet and the other accidents it presents, one concludes demonstratively, as will be seen from what follows, that this planet has either an eccentric or an epicycle. The principle of this demonstration is sense and the sensible effect, that is, the observed motion, as can be seen from the manner in which the *Almagest* ever proceeds: Before it posits the eccentric and epicycle, that book describes, on the basis of numerous observations made at different times and by different astronomers, the movement of the planets. But in addition one encounters certain kinds of strict or mathematical demonstration, for, once the spheres and their movements have been posited, the objects of observation can be inferred demonstratively.

It is obviously this passage in Capuano's *Commentary* to which Nifo was replying in his interpretation of the *De caelo*. And his comeback hits the mark: Though Capuano effectively proved that the Ptolemaic hypotheses are *sufficient* for saving the apparent motions of the planets, he did not demonstrate that they are *necessary*. How, indeed, could he have done so? That would require that he *know* that mankind will never find other assumptions capable of saving the same phenomena.

Nifo's critique clearly shows how foolhardy Capuano was in claiming to prove the truth of Ptolemy's hypotheses.

The Dominican Sylvester of Prierio, instructed, no doubt, by the teachings of Thomas Aquinas, was more judicious. Professor of astronomy at Pavia, he too commented on Purbach's *Theory of the Planets*.[14] This Commentary allows us a glimpse of his opinion on the logical status of astronomical theories. When he describes the shape that Purbach and Regiomontanus ascribed to the sun's orbits, he says:

They do not prove that this is the way things are, and perhaps what they assert is not necessary. . . . The sun, then, has three orbits, that is to say, it is believed that the sun has them; but this is not demonstrated; [the three solar orbits] are thought up solely for the purpose of saving what appears in the heavens.

While the Averroist philosophers on the one side and the Ptolemaic astronomers on the other obstinately persisted in ascribing an inadmis-

14. Sylvester of Prierio of the Order of Preachers, *In novas Georgii Purbachii theoricas planetarum commentaria* (Milan, 1514; Paris, 1515).

sible reality to astronomical hypotheses, the humanists and literati, most of whom had become converts to Platonism, readily endorsed Proclus' opinion on the nature of astronomical hypotheses. Their dilettantism and skepticism also fit in with this way of thinking.

Giovanni Gioviano Pontano, of Ceretto, was born in 1426 and died in 1503. A work of his entitled *De rebus coelestibus libri XIV* was printed for the first time in Naples in 1512 and reissued in the third volume of Pontano's *Opera* in the 1519 edition, which appeared in Venice under the auspices of the Aldes. The work must have been quite popular, since it was reprinted frequently, along with Pontano's other writings. We cite the Basle edition of 1540.[15]

Each of the fourteen books into which Pontano's treatise is divided is preceded by its own *Proemium*, each dedicated to a different person. It is in the *Proemium* to Book III that the author develops, with great clarity and elegance, his conception of astronomical hypotheses.[16]

After recalling that certain astronomers of antiquity attributed the stops and retrograde motions of the planets to the attraction of solar rays, and having taken a stand against any such assumption, Pontano continues as follows:[17]

> The following is, in my opinion, exactly what ought to be believed and thought: These celestial bodies achieve their movements and revolutions spontaneously, in virtue of a power which is properly theirs and without the assistance of external forces, without any kind of attraction by the heat of the sun. Their motions are an achievement of their own nature exclusively.
>
> Nevertheless, those who thought up the ἐπικύκλους (as they are called in Greek) appear to me to deserve the highest praise. It was to find a way of making the senses cooperate with the advance of the understanding that they displayed to our eyes these little circles [the ἐπικύκλοι], to which the planetary bodies are attached and by which they are, while revolving, carried forward or backward, up or down, all in such a way that the true proportions of each movement are preserved. What could be more useful for research, better suited to teaching, than these devices, by which the senses lend their efficacy to the intellect and by means of which the matter that intellectual contemplation pursues is at the same time

15. Giovanni Gioviano Pontano *Librorum omnium, quos soluta oratione composuit, Tomus tertius. In quo Centum Ptolemaei sententiae, a Pontano e Graeco in Latinum translatae, atque expositae;* Pontano *De rebus caelestibus*, bk. 14; Pontano *De luna liber* (incomplete, Basel, 1540; complete, Basel: per haeredes Andreae Cratandri, August, 1540).

16. Pontano *Ad Joannem Pardum de rebus caelestibus liber tertius*, "Prooemium" (pp. 262–76 in the 1540 ed.).

17. Ibid., pp. 267–69.

exhibited to sight? Accordingly, the use of such representations has spread to products of the clockmaker's art which trace out the course of the planets, and to all sorts of little machines and maps—so much so that these devices deserve to be called divine rather than human.

But it would be utterly ridiculous for us to go on and assume that the stars themselves are attached to such circles, that they are transported by them as by chariots.

For, first of all, who would set these circles in motion? Shall we say that they move by their own nature? In that case, why cannot the stellar bodies move of themselves as well? What need is there of outside intervention where the activity of the things themselves suffices? Second, the stellar bodies are visible, since they are formed by the solidification (*concretio*) of the substance of their orbital sphere. But if what is transported is visible because solidification forced it, the circles that do the carrying should also result from solidification, and these rigid circles should likewise be visible.

They are not seen because they do not, in fact, exist. Thought alone sees them, when intent on understanding or teaching. But in the sky there are no such lines and intersections. They have been thought up by extraordinarily ingenious men with a view to teaching and demonstration, since, apart from such a procedure, it would be well-nigh impossible to convey astronomical science, that is, the knowledge of the celestial movements, to others.

The circles, the epicycles, and all suppositions of this sort should, therefore, be regarded as imaginary: they have no real existence in the heavens. They have been invented and imagined so as to let the celestial motions be grasped and to exhibit them to our sight.

The augurs who observe the flight of birds divide the whole of aerial space by means of certain lines. The land measurers distinguish a country into various parts (which they call regions) by means of certain lines running from east to west. . . . Yet neither on the earth nor in the air are there any lines; how much the less on the celestial vault. . . .

Let us defend these suppositions as endowed with a sort of divine virtue as far as teaching and demonstration are concerned. Let us hold that things happen in just this way until, our eyes serving as our guides, we have, by means of a map, learned what we want to know about the stellar motions; until we have grasped them in their numerical values and their dimensions. . . . But as soon as our mind has become thoroughly and correctly steeped in these numbers and magnitudes, as soon as the knowledge of them has penetrated our understanding, let us look upon the circles drawn by astronomers as having far less reality in the sky than the lines traced in the air by the aruspex.

Again:

These spheres are imaginary, for the expanse of the heavens, taken in its entirety, is continuous. Nonetheless, let us retain them as an almost divine invention as long as it is a question of teaching, illustrating, and representing the movement of the stars. Thanks to this invention, the understanding is in possession of a sensible representation that functions as a stepping stone when it begins its investigation. But thereafter the understanding makes headway little by little and ends up by rejecting all such combinations of imaginary spheres to fasten instead solely on numbers and their ratios, which are its proper object.[18]

Pontano's thought is clear: the true goal of astronomy is the numerically exact determination of the celestial movements. Eccentrics and epicycles, and the other hypotheses of astronomy are mere teaching devices, provisional representations which should disappear once astronomical tables and almanacs have been drawn up. Our Renaissance astronomer, unquestionably under the inspiration of Proclus, grants astronomy only two legitimate roles: that of providing the geometric prescriptions requisite to the drawing up of tables which make prediction possible; and that of furnishing mechanical models which enlist the senses in the service of the understanding.

This conception of astronomy, enunciated by Pontano around 1500, was four hundred years later to be regarded as novel.

Excessive confidence in the reality of the objects involved in the hypotheses of astronomy, or exaggerated distrust of the validity of these hypotheses—these are the two extremes between which the Italian philosophers somehow failed to strike the mean. In Paris we shall find thinkers who managed to hold to a more balanced view.

In 1503, Lefèvre d'Etaples published his *Introductorium astronomicum, theorias corporum coelestium duobus libris complectens;* under the title *Fabri stapulensis astronomicum theoricum* the work was reprinted, again in Paris, in 1510, 1515, and 1517; in Cologne in 1516. The lines from the dedicatory epistle quoted below show the spirit in which this treatise was written:

This portion of astronomy is almost entirely a matter of representation and imagination. The good and wise Artisan of all things, by an act of his divine intelligence, produced the real heavens and their real movements. Similarly, our intelligence, which seeks to imitate the Intelligence to which it owes its existence, each day blotting out a little more the spots of its ignorance, our intelligence I say, composes within itself some fictive heavens and fictive motions; these are images of the true heavens and true motions. And in these images, as if they were traces left by the divine Intelligence of the creator, the human intelligence seizes hold of truth.

18. Ibid., p. 273.

When, therefore, the mind of the astronomer composes a correct representation of the heavens and their movements, he resembles the Artisan of all things creating the heavens and their motions.

The hypotheses by means of which Lefèvre d'Etaples represents the celestial motions are, then, as far as he is concerned, not demonstrated propositions. They are not expected to express what the celestial bodies are nor what their true laws of motion are. They are the creatures of the astronomer's genius. By means of these fictions he attempts to provide the imagination with an image of the course of the stars and the planets. They are not truths, these hypotheses, but mere traces, vestiges, images of the truth. God sees the real heavens and their course. The astronomer carries out his geometric constructions and performs his calculations by means of imaginary skies.

These ideas of Lefèvre d'Etaples seems dimly to recall Proclus' thought. Perhaps the influence of Cusa is also to be recognized: Lefèvre d'Etaples greatly admired Nicholas of Cusa and was his disciple; soon after the publication of the *Introductorium astronomicum* he was to edit Cusa's works. Indeed, the characteristics ascribed to astronomical theory by our erudite scientist are much in line with the celebrated Cardinal's characterization of human knowledge in general.

The following principles laid down by Nicholas of Cusa at the beginning of his basic work, the *De docta ignorantia*,[19] are an explanation and a defense of the title of this book.

It is not possible for a finite understanding to appropriate any exact truth. For the true is not something capable of more and less; it is essentially something indivisible, and this something cannot be laid hold of by a being unless this being is truth itself. Similarly, the essence *circle* is something indivisible, and what is not circle cannot assimilate this something to itself. A regular polygon inscribed in a circle is not similar to a circle. It comes to resemble a circle more and more as one increases the number of its sides, but no matter how much one increases that number, a polygon never will become the same as a circle. No figure can be the same as this circle, unless it *be* this very circle.

This is how things stand with respect to truth, and with respect to our understanding, which is not truth itself. Never will our understanding lay hold of truth in so exact a manner that it may not grasp it still more exactly, and it will do so indefinitely.

19. Nicholas of Cusa *De docta ignorantia* 1.1,3.

The true stands, therefore, in some kind of opposition to our reason. *Truth* is a necessity which admits neither of diminution nor of increase; whereas *Reason* is a possibility ever susceptible of new development. Of the true, then, we know nothing, except that we cannot comprehend it.

What conclusion ought we to draw from this?

That the very essence of things, that which is the true nature of beings, cannot ever be reached in its purity, not by us. All philosophers have sought it; none have found it. The more profoundly learned we become in this ignorance, the closer we approach to truth itself.

What perfection, then, should the student strive for? To become as learned as is possible in this ignorance, which is his proper estate: "He will be the more learned, the more he knows himself as ignorant."

Proclus had distinguished two kinds of physics: the one intent upon knowing the essence and the causes of sublunary things, a physics accessible to man; the other having the nature of celestial things for its object, a physics reserved for the divine Understanding.

Nicholas of Cusa regarded the stars as of the same nature as the four elements. For him, therefore, the distinction set up by Proclus loses all sense. Yet he continues to discriminate between two kinds of physics, though he contrasts them with each other in an entirely new way.

One physics is the knowledge of essences and causes. It meets the requirements of the definition that scholastic philosophy had imposed on all knowing: *scire per causas*. Necessarily perfect and immutable, it is not accessible to man but is God's science.

The other physics is of a radically different sort: they are as heterogeneous as polygon and circle. It does not know genuine causes and essences. If it uses these words, it can apply them only to hypothetical causes and fictive essences, which are creatures of reason, not realities. The physics so constituted is ever on the way of self-perfection. The physics of essences and causes functions as its limit, giving direction to its development. Yet it is forever barred from reaching its limit. The physics of fictions and abstractions is the only physics accessible to man.

The opposition between physics and astronomy that the Greek thinkers —Posidonius, Ptolemy, Proclus, Simplicius—had established is displaced by Nicholas of Cusa. In its stead is the opposition between the absolute physics of real essences and genuine causes and the relative and developing physics of abstract essences and fictive causes.

When the Spaniard Luiz Coronel was writing his work on physics at the Collège de Montaigu, in 1511,[20] was he under the influence of Nicholas of Cusa? It is entirely possible that he was. The works of the German Cardinal were well known at the time. They had been printed twice by then, and in 1514 Lefèvre d'Etaples was to prepare a third edition to be published in Paris! However this may be, certain opinions that the Rector of the Collège de Montaigu defends in his *Physicae perscrutationes* agree very closely with the principles of the *Docta ignorantia.*

For Luiz Coronel, physics is not a deductive science whose propositions follow from principles evident a priori. It is a science whose origin lies with experience, and the principles of cosmology are nothing but hypotheses conceived with a view to saving the phenomena that experience has made known to us.

When, for example, he proposes to establish that there is not only a form in every substance, but also a material,[21] Coronel adopts the following experimental fact as his point of departure: we cannot start a fire without destroying something combustible. The notion of matter is here shown indispensable by virtue of the fact that the phenomenon could not be accounted for if fire were pure form. Generalizing the method he has just employed, he ventured the following axiom: "In physics, arguments drawn from experience should be given primacy" (*rationes ex experientia sumptae in physica obtinent primatum*).

This is how he defends their title to primacy:

As Albertus Magnus maintained, in the discipline of physics arguments drawn from experience fill the role of principles (*rationes ex experientia sumptae in physica disciplina obtinent principatum*). The arguments of the astronomers, drawn from the diversity of the celestial motions and the distances between the heavenly bodies and the planets, led to the proposal of epicycles, eccentrics, and deferents as conclusions. Similarly, matter must be posited, as required by natural reason. For if it is not, the fact that making a fire requires the supply of something combustible cannot be saved (*non potest salvari*), just as the celestial appearances cannot be saved unless one posits epicycles etc. The hypothesis of epicycles, eccentrics, and deferents offended the Commentator, Averroes, but he provided no alternative method of saving what is saved by these assumptions. Moreover, the same might be said of matter and the other causes which he himself admitted to save the things that occur naturally (*et sic dicere-*

20. Luiz Coronel, *Physice percrustationes.* Prostant in edibus Joannis Barbier librarii jurati Parrhisiensis Academie sub signo ensis in via ad divum Jacobum (1511).

21. Ibid., fol. 2, col. a.

tur etiam ei de assignatione materiae et aliarum causarum naturalium quas ipse ponit ad salvandum ea que naturaliter contingunt).

To account for Luiz Coronel's point of view we do not have to appeal to the influence of Nicholas of Cusa. It would be sufficient to invoke the traditions of the University of Paris; Coronel was merely formulating a rule of procedure constantly observed at that university from the middle of the fourteenth century on, as can be seen from the works of John Buridan, Albert of Saxony, and Nicholas of Oresme, which supply many examples.

For one such example, let us turn to a happy theory of John Buridan, tenaciously defended by him: A projectile's movement is not, as Aristotle would have it, maintained by the motion of the surrounding air; rather, it is due to a certain quality or *impetus* engendered in the substance of the projectile by whoever hurled the body. Having shown that all the other hypotheses previously proposed by sundry philosophers are in various ways contradicted by experience, Buridan says of his own theory:

> It seems to me that this is the assumption that should be adopted because the other assumptions do not seem correct and because all the phenomena agree with this one (*hujusmodi etiam modo omnia apparentia consonant*).[22]

All the facts of experience known to him are brought to bear on his hypothesis. Is not the method that John Buridan here follows the very method Luiz Coronel was preaching?

Upon collating the ideas of Coronel, John of Jandun, and Lefèvre d'Etaples, the following conclusion is, we feel, warranted: between 1300 and 1500 the University of Paris taught a doctrine of physical method which far surpassed in truth and profundity all that was going to be said on this subject until the middle of the nineteenth century.

One powerful and fruitful principle promulgated and observed by the Parisian Scholastics deserves special mention—the recognition that the physics of the sublunary world is not heteregeneous with celestial physics, that both proceed by one and the same method, that the hypotheses of the former as well as those of the latter are geared toward a single end—*to save the phenomena.*

22. Paris, Bibliothèque nationale, fonds latin, ms. no. 14723, fol. 106, col. d (John Buridan "Quaestiones totius libri phisicorum" 8.12).

5
Copernicus and Rheticus

This limpid conception of the nature of physical hypotheses of several medieval and early Renaissance thinkers became gradually obscured; in the centuries to follow it received its greatest set-back precisely at the time when astronomy and physics were making new and rapid progress. The greatest artists are not necessarily best at philosophizing about their art.

On May 24, 1543, Copernicus died. That same year his immortal masterpiece *On the Revolutions of the Celestial Spheres* was printed.[1] In the dedicatory letter he addressed to Pope Paul III, Copernicus explains the general attitude and direction of his thought:

> What Your Holiness chiefly expects from me is to learn how the bold fancy of ascribing a certain movement to the earth entered my mind, in spite of the established opinion of mathematicians and almost in violation of common sense. I want Your Holiness to understand the one and only motive that led me to conceive of a new reason for the movements of the celestial spheres. It is this: I saw that mathematicians disagree among themselves about the investigation of these movements. First of all, they have remained in such uncertainty about the movements of the sun and moon that even today they are unable either to observe or to prove what is the invariable length of the year. Second, when they want to construct the movements of these two stars and of the five planets, they do not set out from the same principles, nor from the same hypotheses; they do not

1. Copernicus *De revolutionibus orbium coelestium*, bk. 6 (Nuremberg: apud Joh. Petreijum, 1543).

explain the apparent revolutions and movements in the same way, for some use only homocentrics while others use eccentrics and epicycles; and even so they do not fully meet the requirements of astronomy. Those who trust to homocentrics do prove that certain irregular movements can be constituted by this procedure; but on their hypotheses they have not been able to establish anything precise that corresponds exactly with phenomena. Those who imagine eccentrics seem thereby to have analyzed most of the apparent movements in a way that makes them agree numerically with the tables; but the hypotheses they have accepted appear for the most part to contravene the prime principles of equality of movement. Furthermore, they have been unable to discover or deduce from their assumptions the one thing which is most important, namely, the shape of the world and the exact symmetry of its parts. . . . In the course of demonstration—*μέθοδον* as it is called—they have obviously either left out some necessary condition or introduced some foreign assumption extraneous to the subject. This would surely not have happened to them if they had been following principles that are certain. If the hypotheses they adopted were not mistaken, everything that follows from them would doubtless have been verified.

The text just quoted puts us in mind of the great debates that stirred the Italian universities when Copernicus came to Italy to study: discussions about the reform of the calendar and about the theory of the procession of the equinoxes; and the bitter quarrel between the Averroists and the partisans of the Ptolemaic system. It was the friction between these two schools that fired the spark that set the genius of Copernicus aflame.

Copernicus thought of the astronomical problem in the manner of the Italian physicists whose courses he had audited or who had been his fellow-students. The problem was *to save the appearances by means of hypotheses conformable to the principles of physics.* When it is formulated in this way, neither the Averroists nor the Ptolemaists solve the problem adequately: The former adopt hypotheses which are physically tenable but do not save the appearances; the latter save the appearances rather well but their assumptions contravene the principles of the science of nature. If both parties are unable to furnish the solution expected of them, this must surely mean that their hypotheses are false. A fully satisfactory astronomy can only be constructed on the basis of hypotheses that are *true*, that conform to the nature of things.

Copernicus undertook to look for these true hypotheses.

After long mulling over this uncertainty of the mathematical traditions concerning the theory of the celestial movements, I was overcome by disappointment that philosophers who had so minutely examined the least things of this world should not have found any more certain reason for the movement of the world machine.[2]

Driven by this disappointment, Copernicus—who was a humanist—searched through the works of Greek and Latin authors. From Cicero and the author of *De placitis philosophorum* he learned that several ancient thinkers had set the earth in motion:

Under the influence of this suggestion, I began myself to consider the movement of the earth. It seemed an absurd notion. Yet I knew that my predecessor had been granted the liberty to imagine all sorts of fictive circles to save the celestial phenomena. I therefore thought that I would similarly be granted the right to experiment, to try out whether, by assigning a certain movement to the earth, I might be able to find more solid demonstrations of the revolutions of the celestial spheres than those left by my predecessors.

Now, in fact, long and repeated observation has shown me that, by assigning the various movements to the earth that I ascribe to it later in this work, all the other wandering stars' appearances follow from a computation by which the motions of the stars, taking each one into account, are referred back to the earth; it has shown me besides that on this assumption the order and the magnitudes of the stars, of the various spheres, and even of the sky itself, turn out so intimately bound up with one another that it becomes impossible to rearrange anything in any portion of the sky without thereby throwing all the other parts, and the whole, into confusion.

At first, Copernicus tested the hypothesis of a moving earth as a purely fictive assumption, and he found that on this assumption the phenomena were saved. Was it sufficient for his purposes to have established this much? The final sentence in the last passage quoted shows that he wanted to do more, that he was anxious to prove the truth of his hypothesis, and that he thought he had succeeded in doing so. To demonstrate that an astronomical hypothesis is in conformity with the nature of things, more is required than to show that it is sufficient for saving the phenomena; one must prove besides that these phenomena could not be saved if the hypothesis were rejected or modified. Nifo had rightly insisted on the indispensability of this supplement. It would appear that Copernicus fell victim to an illusion similar to the one that had let the Ptolemaic astronomer Capuano of Manfredonia astray, that Copernicus too ascribed a value to his system which could only have been conferred upon it by such a supplementary proof of the system's necessity.

2. Ibid. "Ad Sanctissimum Dominum Paulum III Pontificem Maximum, Nicolai Copernici praefatio in libros Revolutionum."

Copernicus' own work hardly more than implies this larger claim. But the little treatise which Joachim Rheticus composed in 1540 is quite explicit on the point.[3] Rheticus here provides a summary preview of the doctrines his master had so long delayed publishing. In the *Narratio prima* we first find the idea sketched somewhat hesitantly; gradually it becomes more firmly drawn; and finally it is set down quite definitely:

You [Rheticus is addressing Schoner] are well aware what place hypotheses or theories have in astronomy, and how much the mathematician differs from the physicist. You will therefore agree, I feel, that we must go where observation and the very testimony of the sky itself lead us.[4]

What instruction can we expect to receive from these observations, from what Rheticus calls "the testimony of the sky itself"? Should we, following the method of Aristotle, require that they provide us with knowledge of the efficient causes of phenomena? Or should we, following Ptolemy, ask merely that they suggest fictive hypotheses suitable for saving these same phenomena? Rheticus mentions only the first of these two alternatives:

Aristotle confirms, by his own example and by that of Calippus, that astronomy's proper goal is to assign the proper causes τῶν φαινομένων and to do this in such a way that the various motions of the celestial bodies result from these causes.

He holds that his master's doctrine meets not only the requirements of Ptolemy but those of Aristotle as well:

In astronomy, just as in physics, one usually moves from the effects and from observation to the principles. I am convinced that Aristotle, once he understood the grounds for the new hypotheses, would candidly acknowledge which things in his discussions of the heavy and the light, of circular motion, and of the earth's rest and motion were demonstrated and which were laid down as principles without being demonstrated.

3. *Ad clarissimum virum D. Joan. Schonerum de libris revolutionum eruditissimi viri et mathemaci excellentissimi Reverendi D. Doctoris Nicolai Copernici Torunnaei, Canonici Varmensis, per Quendam Juvenem Mathematicae Studiosum, Narratio prima* (Gedenum, 1540). We shall cite Rheticus' *Narratio prima* after the first edition prepared in commemoration of the fourth centennial of the birth of Copernicus: *Nicolai Copernici Thorunensis De revolutionibus orbium caelestium libri VI. Ex auctoris authographo recudi curavit Societas Copernicana Thorunensis. Accedit Ioachimi Rhetici De libris revolutionum narratio prima.* (Thorn: sumptibus Societatis Copernicanae 1873).

4. Rheticus *Narratio prima, transitio ad enumerationem novarum hypothesium totius Astronomiae* (pp. 463–64 in the 1873 ed.).

Rheticus, we see, believed that his master, in devising his new hypotheses, was not merely doing geometer's work but also physics. In his opinion, Copernicus had set up a new physics that was destined to supplant the ancient Peripatetic physics, a physics that Aristotle, were he alive, would have endorsed.

Copernicus arrived at his hypotheses by the method of the physicist, that is, by moving from effects to causes. To what degree of certainty could he thus attain? Rheticus tells us:

Aristotle says: "That is most true which is the cause of the truth of what follows from it" (*verissimum est id quod posterioribus, ut vera sint, causa est*). My master therefore believed that he should lay down such hypotheses as contain within themselves causes capable of confirming the truth of observations made in earlier times and as would, in addition, give reason to hope that they might, in the future, be causes of the truth of all astronomical predictions τῶν φαινομένων.[5]

The conclusion, not explicitly drawn by Copernicus' faithful disciple, is inescapable: The Copernican hypotheses are, to use the Latin rendering of Aristotle, *verissimae,* "most true."

Indeed, Rheticus is so convinced of the adequacy of the hypotheses to the phenomena that he regards them as mutually interchangeable, like *definiens* and *definiendum:*

I hope that both accounts (Narrationes) will be all the more acceptable to you, the more clearly you perceive that in view of the observations made by scholars, the hypotheses of my learned teacher correspond so well to the phenomena that they can be mutually interchanged, like a good definition with the thing defined.[6]

Faithful disciple of Copernicus, Rheticus is neither Averroist nor Ptolemaist, yet he cherishes the same ideal of astronomical theory as did the Ptolemaist Capuano or the Averroist Nifo. He too holds that a good astronomical system does not only save the celestial phenomena and enable one to calculate the stellar motions exactly, but is, besides, a system built on hypotheses that are founded in the very nature of things.

5. Ibid., "Universi distributio" (p. 464).

6. "Et vero gratiorem tibi utramque Narrationem fore spero, quo clarius artificum propositis observationibus ita D. Praeceptoris mei hypotheses τοῖς φαινομένοις consentire videbis, ut etiam inter se tanquam bona definitio cum definiti converti possint," Rheticus *Narratio prima,* "Quomodo planetae ab ecliptica discedere appareant," (p. 489 in the 1873 ed.).

6

From Osiander's Preface to the Gregorian Reform of the Calendar

When it was published at last, the book in which Copernicus expounded his astronomical theory contained views about the hypotheses supporting such a theory that were absolutely opposed to the views that would seem to have inspired Copernicus and Rheticus. The book opens with an unsigned preface bearing the title: *Ad lectorem, de hypothesibus hujus operis.* This preface reads as follows:

Since the novelty of the hypothesis here proposed, according to which the earth moves and the sun stays fixed at the center of the universe, has already received a great deal of publicity, I have no doubt but that certain of the savants have taken grave offense and think it wrong to upset the liberal disciplines which have so long been firmly established. If, however, they are willing to weigh the matter scrupulously, they will find that the author of this book has not done anything deserving of censure.

For the astronomer's job consists of the following: To gather together the history of the celestial movements by means of painstakingly and skilfully made observations, and then—since he cannot by any line of reasoning reach the true causes of these movements—to think up or construct whatever hypotheses he pleases such that, on their assumption, the selfsame movements, past and future both, can be calculated by means of the principles of geometry. . . . It is not necessary that these hypotheses be true. They need not even be likely. This one thing suffices that the calculation to which they lead agree with the result of observation (*neque enim necesse est, eas hypotheses esse veras, imo ne verisimiles quidem, sed sufficiet hoc unum, si calculum observationibus congruentem exhibeant*). . . . Obviously this science simply does not know the causes of the irregularity of the apparent movements. It thinks up fictive causes which, generally speaking, it views as known with certainty; yet it is not with an eye to ever persuading anyone that this is how things really are that it so conceives of the hypotheses, but solely to set up correct computation. Sometimes alternative hypotheses are available with which to account for one and the same movement; the eccentric and epicycle in the theory of solar motion are a case in point. In such a case, the astronomer will by preference choose the hypothesis that is easier to grasp while the philosopher tends to seek out likelihood. Neither the one nor the other can, however, either conceive or enunciate the least certainty, unless he be the recipient of some divine revelation. . . . Let no one, then, expect from astronomy any doctrine about these hypotheses that is certain. Astronomy can give him nothing of the sort. Let him take care not to take as true, assumptions which were fabricated for quite a different purpose, lest, far from gaining access to astronomical science, he be turned away from it, and leave it more stupid than he was before.

Nicolas Müller, who in 1617 put out the third edition of Copernicus' book, likens the just-quoted suggestions to those we find in the *Almagest.*[1] He might with equal justice have compared them to a good many others, for the opinion that is so clearly stated in the unsigned preface to the *De revolutionibus* is the echo of the Greek tradition which, through Geminus, Ptolemy, and Proclus, stretches from Posidonius to Simplicius; it is the echo of the critique of Maimonides; also of the tradition of Paris that was born of the teaching of Thomas Aquinas and Bonaventura and handed on by John of Jandun and Lefèvre d'Etaples. In a word, the anonymous preface is the echo of that tradition in the history of astronomy which protested without letup against the realism of thinkers like Adrastus of Aphrodisias and Theon of Smyrna, the Arab physicists, the Italian Averroists and Ptolemaists, Copernicus and Rheticus themselves.

Who wrote this preface? Kepler tells us.

In 1597, Nicolas Ryemer Baer (Raimarus Ursus) published a work in which he meant to defend opinions analogous to those contained in the preface.[2] Kepler wanted to answer Ryemer. So, three years after the pub-

1. Copernicus *Astronomia instaurata, libris sex comprehensa, qui de Revolutionibus orbium coelestium inscribuntur.* Nunc demum post 75 ab obitu authoris annum integritati suae restituta. Notisque illustrata, opera et studio D. Nicolai Mulerii Medicinae ac Matheseos Professoris ordinarii in nova Academia quae est Groningae. (Amsterdam: excudebat Wilhelmus Iansonius, sub Solari aureo, 1617).

2. Nicolas Ryemer Baer *Tractatus astronomicus de hypothesibus astronomicis, seu systemate mundano; item, astronomicarum hypothesium a se inventarum, oblatarum et editarum, contra quosdam, eas temerario ausu arrogantes, vindicatio et defensio, cum novis quibusdam subtilissimisque compendiis et artificiis in nova doctrina sinuum et triangulorum.* (Prague, 1597.)

lication of Ryemer's treatise, that is, around 1601, he composed a violent lampoon.[3] This pamphlet was not published until Ch. Frisch discovered an incomplete copy of it among Kepler's papers.[4] We shall shortly return to this debate between Ryemer and Kepler. For the moment we want to take up only one point.

Ryemer did not know who the author of the *Prefatio ad lectorem* was. Says Kepler:[5]

I shall come to Ursus' rescue. The author of this preface is Andreas Osiander, as a note on the copy I possess testifies—a note in the handwriting of Jerome Schreiber, to whom Schoner addressed several of his prefaces.

Kepler goes on to explain that the preface added to the *De revolutionibus* after its author's death expresses neither Copernicus' own thought, nor even the real thought of Osiander,[6] who wrote it. Copernicus himself candidly revealed his attitude to the hypotheses underlying his book. But Osiander, fearing that the apparent absurdity of the Copernican hypotheses might alarm the philosophizing crowd (*vulgus philosophantium*), considered it advisable to minimize the scandal he anticipated: thence the idea of placing the notorious preface at the head of the book *On the Revolutions.* In support of these assertions, Kepler cites two letters of Osiander. On April 20, 1541, Osiander writes to Copernicus:

As for hypotheses, this is what I have always thought on that subject: they are not articles of faith, they are merely the basis of calculation; even if they should be false, that hardly matters, so long as they reproduce the φαινόμενα of the movements exactly (*de hypothesibus ego sic sensi semper, non esse articulos fidei, sed fundamenta calculi, ita ut etiamsi falsae sint, modo motuum φαινόμενα exacte exhibeant, nihil referat*).

For consider, if we follow Ptolemy's hypotheses, who can assure us whether the irregular movement of the sun occurs rather in virtue of the epicycle or in virtue of the eccentric, since it can be produced in either way? I would urge you to touch on this question in your preface; you would thereby pacify the Peripatetics and theologians whose opposition you fear.

That same day he writes to Rheticus:

The Peripatetics and theologians will easily be placated if they are made to understand that different hypotheses may correspond to one and the

3. Johannes Kepler *Apologia Tychonis contra Nicolaum Raymarum Ursum.*

4. Kepler *Opera omnia*, ed. Frisch (Frankfurt and Erlangen, 1858), vol. 1, p. 215.

5. Ibid., p. 245.

6. Andreas Hossmann, who "hellenized" his name, in accordance with the fashion of his day, to Osiander.

same apparent motion; that they are not advanced as expressing the real with certainty but rather as guiding the computation of the apparent and composite motions most conveniently; that different authors may think up different hypotheses; that one may propose a representation which is quite well suited, and another a representation which is still better suited, both meanwhile nevertheless engendering one and the same apparent motion; that everyone should be left free to seek hypotheses more convenient than those hitherto accepted; that we should even be grateful to anyone who makes efforts in this direction. . . .

By these quotations, valuable though they are, Kepler has proved only part of what he claimed: We see quite clearly that Osiander, by putting his famous preface at the head of the treatise *On the Celestial Revolutions*, was going counter to the realist intentions of Copernicus and Rheticus; reading their works has already convinced us of this much. But it is not at all obvious that Osiander, by a subterfuge designed to pull the wool over the eyes of Peripatetics and theologians, was dissimulating his own thought. On the contrary, it seems quite clear from Osiander's letter to Copernicus that he had for long been convinced of the *truth* of the doctrine which two years later he was to publish in his celebrated *praefatio ad lectorem.* He notes, quite correctly, that this doctrine nullifies every objection raised in the name of cosmology or revelation against this or that system of hypotheses. But there is nothing to justify the view that he is hiding his own convictions to gain this tactical advantage.

Osiander is by no means alone in this attitude toward astronomical hypotheses in general and those of Copernicus in particular.

In 1541, Gemma Frisius, the celebrated Dutch astronomer, writes a letter to Dantiscus, from Louvain, in which he speaks of Copernicus in the following terms:

I won't become embroiled in any argument about the hypotheses he uses for his demonstration; I don't investigate what they are nor what portion of truth they contain. It hardly matters to me whether he claims that the earth moves or declares it immobile, so long as we get an absolutely exact knowledge of the movements of the stars and of the periods of their movements and so long as both are reduced to altogether exact calculation.[7]

In fact, Osiander was merely applying to the Copernican hypotheses what the Ptolemaists had over and over again said about the hypotheses of the *Almagest:* these hypotheses were close to their hearts in that they facilitated the construction of astronomical tables and almanacs—construc-

7. Cited by Leopold Prowe, *Nicolaus Copernicus* (Berlin, 1883), vol. 1, pt. 2, p. 184.

tions to which the disciples of Purbach and Regiomontanus devoted painstaking care; but most of them held the reality of hypotheses as cheaply as did Osiander.

Of this state of mind we find the most telling example in the writings of Erasmus Reinhold.

Erasmus Reinhold of Saalfeld wrote his first work, a commentary on Georg Purbach's *Theories of the Planets,*[8] in Wittenberg. It appeared there in 1542, with some distichs and a preface by Melanchthon. The preface, dated 1535, was, it seems, originally intended for an earlier edition of Purbach's *Theories*, an edition for which Jacob Melichius had designed the illustrations.

The *Theories of the Planets* by Purbach, as Melanchthon notes in his preface, is an introduction to the astronomical system of the *Almagest.* It presents this system in synthetic and deductive form: The Ptolemaic hypotheses are laid down at one swoop and the explanation of phenomena is then deduced from them in geometric fashion. Melanchthon characterizes this method by saying that Purbach proceeds according to τὸ ὅτι. This method is contrary to the one followed by Ptolemy, who proceeded analytically and inductively: He discussed the phenomena and, through such discussion, suggested what hypotheses might enable one to represent the phenomena. Reinhold characterizes this order of exposition by the word διότι, contrasting it with the order according to τὸ ὅτι. Thus, speaking of the theory of the moon, he writes:

> You see the διότι of this portion of astronomy, and with what subtlety, what skill, Ptolemy uses observations in going after the causes of the φαινόμενα.

What does Reinhold mean here in saying that Ptolemy investigated the *causes* of *phenomena*? Is he talking about efficient causes in the metaphysical sense of the word? By no means! *Causes* of celestial *phenomena*, an expression frequently encountered in the writings of Reinhold and his contemporaries, means no more and no less than this: the simple motions from whose composition the apparent motions are engendered. Ptolemy

8. *Theoricae novae planetarum Georgii Purbachii Germani ab Erasmo Reinholdo Salveldensi pluribus figuris auctae, et illustratae scholiis, quibus studiosi praeparentur, ac invitentur ad lectionem ipsius Ptolomaei. Inserta item methodica tractatio de illuminatione Lunae* (1542). In fine: Impressus hic theoricarum libellus Vitembergae per Ioannem Lufft (1542). This edition was reprinted without change in 1556, 1557, 1558, "Parisiis, apud Carolum Perier, in vice Bellovaco, sub Bellerophonte."

proceeded according to the διότι (the "whence" or "whereby") because he went back from the phenomena to their causes, that is to say, he studied the apparent motions so as to discover from what combinations of simple motions they might result. Purbach, by contrast, proceeds according to the ὅτι in that he takes the combinations of simple motions for granted and deduces the properties of the apparent motions from them.

That this interpretation is correct becomes perfectly plain when we look at Reinhold's own preface to his Commentary:

> The variety of the celestial movements and appearances (which the Greeks call φαινόμενα) is overwhelming. So astronomers have been extremely meticulous, spent many sleepless nights, and devoted much wearisome labor to the investigation of the causes of these very varied appearances. . . . To make the causes of the variegated appearances shown by the planetary motions known, learned astronomers have, speaking generally, assumed or established either the eccentricity of the deferents or the multiplicity of the spheres. The numerousness of the spheres thus obtained must be attributed to the astronomer's art, or rather, to the weakness of our understanding. Perhaps these seven shining, beautiful stars have in themselves a certain power, given by God, whereby each can follow its own law without requiring the assistance of any such spheres, whereby each preserves a perpetual harmony throughout the variety and apparent irregularity of its movement. But for us, if we did not call upon these spheres for aid, it would be extremely difficult to get a rational hold of this sort of order in disorder; we would not be able to keep it in mind and follow it up in thought.

Reinhold's ideas here tie in with those of Proclus: There is no real motion over and above the complicated, unanalyzed motion that the several stars exhibit. The revolutions along eccentrics and epicycles into which Ptolemy's astronomy had resolved these motions are merely contrivances designed to facilitate our study of the former.

But if, as pure constructs, these component motions have no reality, if they are simply vehicles for reasoning and calculation, they are essentially variable and perfectible: For the "causes" that Ptolemy proposed, other "causes" that save the phenomena more exactly or more conveniently may be substituted. Reinhold could not be prevailed upon by physical arguments to reject the transformation to which Copernicus would subject astronomical hypotheses.

Not only did he not reject this transformation—the main lines of which were already known to him, no doubt through Rheticus' *Narratio prima*—he awaited it with impatient curiosity.

Thus, in his preface, having roused the reader's admiration for the ingenuity shown by Ptolemy in his construction of the theory of the moon, Reinhold adds:

I know of a modern scientist who is exceptionally skilful (*quendam recentiorem praestantissimum artificem*). He has raised a lively expectancy in everybody. One hopes that he will restore astronomy. He is just about to publish his work. In the explanation of the phases of the moon he abandons the form that was adopted by Ptolemy. He assigns an epicyclic epicycle to the moon. . . .

Later, when he is about to deal with the precession of the equinoxes, Reinhold writes:

For long these sciences have been waiting for a new Ptolemy capable of putting these studies on their feet and of setting them off again on the right road. I hope that this astronomer, whose genius all posterity will rightly admire, will at long last come to us from Prussia. . . .[9]

One year after the publication of these words the *De revolutionibus orbium coelestium libri sex*, so ardently anticipated by Reinhold, made its appearance. As soon as he had become familiar with the new methods by which Copernicus provided "causes" for the celestial "phenomena," Reinhold took as great an interest in them as he had earlier taken in the doctrines of Ptolemy. He wrote a commentary on the Copernican system; as far as we know, it was never published. What is more, he undertook to provide as indispensable supplement to the work of Copernicus by drawing up new astronomical tables based on the theories proposed by Copernicus. In 1551 he published these *Prutenicae tabulae,*[10] which greatly contributed to extending the use of Copernicus' theories among astronomers.

From the enthusiastic outpouring in the introduction to the *Prutenicae tabulae* it is clear that Reinhold's expectations were in no way disappointed when the *De revolutionibus* was finally published. His admiration for the inventions of the astronomer of Thorn is expressed in statements like:

All posterity will gratefully celebrate the name of Copernicus. The science of the celestial motions was almost in ruins; the studies and works of

9. Ibid., "De motu octavae sphaerae" (toward the end of the preface).
10. Reinhold *Prutenicae tabulae coelestium motuum* (Wittenberg, 1551).

this author have restored it. God in his goodness kindled a great light in him so that he discovered and explained a host of things which, until our day, had not been known or had been veiled in darkness.[11]

It is not hard to believe that the author of this eulogy, previously a convinced Ptolemaist, had become an enthusiastic Copernican. And the title of faithful disciple of Copernicus should, indeed, not be withheld from him if one takes the conversion to mean that the astronomer of Wittenberg admired the simplicity and ease of the geometric constructions proposed by the new system, that he believed them better adapted to calculation than the combinations of the *Mathematical Syntaxis.* But it would, we think, be foolhardy to conclude, from this that Reinhold really believed in the motion of the earth and the fixity of the sun. The *Tabulae prutenicae* seem to treat these hypotheses simply as geometric devices for the construction of astronomical tables, devices similar in nature to the Ptolemaic ones. Consider, for example, the following passage:

It should be known that the diurnal movement of a planet is the sum of two parts: the first is the true movement of the epicycle, which Copernicus sometimes calls the earth's movement and sometimes apparent motion (*quem Copernicus alias Terrae, alias visum motum . . . nominat*); the other is the true movement of the planet by which it is itself animated, for instance, according to the customary Ptolemaian hypotheses, the movement along the circumference of an epicycle.

In Reinhold's entire book there is but *one word* that might be taken to suggest that the author ascribes some sort of reality to astronomical hypotheses: The *Logistice scrupulorum astronomicorum,* which serves as an introduction to his *Prutenicae tabulae*, opens with a preface, and in this preface we read:

Astronomy cannot be established and completed except with the help of two sciences which are, as it were, its instruments—namely, geometry and arithmetic. . . . Geometry plays a double role in the constitution of astronomy: in the first place, it offers hypotheses that are in agreement with the anomalies of the revolutions; second, in order that the science, reduced to numbers, can easily and always be employed for the purposes of daily life, it provides us with that masterful and thorough-going method of calculation called trigonometry. . . . So geometry is in charge of both parts of astronomy—of Θεωρητικὴ, which subordinates the study

11. Reinhold *Logistice scrupulorum astronomicorum* (Wittenberg, 1551). "Praecepta calculi motuum coelestium" 21.

of movements to hypotheses that are certain (*certis hypothesibus*), and of Ποιητικὴ, which with marvelous skill and ingenuity reduces the stellar movements to numerical tables or, by means of these, to exact instruments (*certa instrumenta*).[12]

How are we supposed to construe the words *certae hypotheses?* Should we interpret them as synonymous with "physically true hypotheses"? Are "certain hypotheses" hypotheses that "conform to the nature of things"? Reinhold's entire book seems to cry out against this interpretation. Moreover, since only a few lines later we are obliged to render *certa instrumenta* as "exact instruments," is it not plausible to take *certae hypotheses* likewise simply as "exact hypotheses"? In short, everything induces us to look upon Reinhold as a scientist who, with regard to astronomical hypotheses, follows the views of Osiander.

Osiander, Frisius, and Reinhold were, of course, not the only astronomers of their day to think thus. Reinhold, in particular, taught at the University of Wittenberg, together with Melanchthon. And under the influence of these two teachers there was formed a circle of disciples who shared their professors' opinions on the nature of astronomical hypotheses.

Ariel Bicard was such a disciple and admirer of Reinhold. In his *Questions concerning John of Sacro-Bosco's Treatise on the Sphere*[13] he briefly touches on the theory of eccentrics and epicycles and, in this connection, raises the question: "Are these planetary orbs real?"[14] His answer, as succinct as it is unequivocal, is: "Such orbits do not really exist in the heavens; we merely imagine them to help those who are learning astronomy (*propter discentes*), so that, in this way, the movements of the heavenly bodies may be saved."

Kaspar Peucer, like Ariel Bicard, was a student at the school of Wittenberg headed by Melanchthon and Reinhold. At the beginning of the *Elements of the Doctrine of the Celestial Circles,* which he published

12. Ibid. 36.

13. *Quaestiones novae in libellum de Sphaera Joannis de Sacro Bosco, in gratiam studiosae iuventutis collectae ab Ariele Bicardo, et nunc denuo recognitae, figuris mathematicis ac tabulis illustratae, quae in reliquis editinibus antehac desiderabantur* (Paris: apud Gulielmim Cavellat, in pingui gallina, ex adverso collegii Cameracensis, 1552). The preface, surely contemporaneous with the first edition, is dated 1549.

14. Ibid. "Quaestiones in quartum librum Sphaerae" (first part of the book), 1, "De numero orbium Solis," fol. 70 (verso).

in 1551 and reprinted in 1553,[15] there is a poem in which we find Melanchthon decorated by Peucer with the title of "Father." The work starts out with a chronological list of astronomers, extending from the creation of the world up to the year A.D. 1550. The last name mentioned on the list is Erasmus Reinhold's, whom Peucer describes as *"praeceptor mihi carissimus et perpetua gratitudine celebrandus."*

Not surprisingly, Peucer talks about astronomical hypotheses in much the same way as Reinhold.

Peucer's *Elements* are almost exclusively devoted to the study of the "first" or diurnal movement. The book's plan is nearly identical with that of John of Sacro-Bosco's *On the Sphere.* The study of the "secondary mobiles," that is, the wandering stars, is relegated to the *Theory of the Planets.* Only in the last two pages of the *Elements* does Peucer touch on it with a few words:

> The secondary mobiles and the secondary movements which are the movements of the eighth sphere and of the seven wandering stars, display great differences and manifold variety, as is shown by the φαινόμενα and observation of these phenomena.
>
> Observation has convinced us that this diversity and variety is to be met with in the movement of every one of these stars, but it likewise teaches us that their movements recur according to a fixed and immutable law. It is therefore quite certain that each sphere's movement has a certain period at the end of which the revolution has become complete. To leave nothing irregular in the heavenly movements, some astronomers save these φαινόμενα by means of certain hypotheses, others save them by means of other hypotheses (*salvant haec alii aliis hypothesibus constitutis*); they admit eccentrics and epicycles, some more, some fewer. Starting from these hypotheses, they construct proofs by means of which they show the causes of the variety in question. The treatises, *Theories*, . . . explain and develop these hypotheses.

Surely, here speaks a man who does not view astronomical hypotheses as expressing realities but takes them merely as constructs designed to save the phenomena.

Moreover, even though Peucer in the passage we have just cited alludes only to hypotheses patterned after the Ptolemaic system, he would probably have been just as willing to accept Copernicus' hypotheses if they

15. Kaspar Peucer *Elementa doctrinae de circulis coelestibus, et primo motu, recognita et correcta.* (Wittenberg: excudebat Jiohannes Crato, 1558). The first edition, which we were unable to consult, is dated 1551.

saved the *φαινόμενα* more exactly. This attitude alone already explains the admiration he professes for Copernicus, an admiration measured by the prominence given to the least detail of Copernicus' life in the chronological list we mentioned. Thus Peucer is careful to tell us that:

Nicholas Copernicus of Thorn, Canon of Varmia, was born in the year 1473, on the 19th of February, at forty-eight minutes after four.

And he enters the name of the quite mediocre Domenico Maria Novara of Bologna on his list with the remark, "of whom Copernicus was a student and assistant." Furthermore, Peucer borrows from Copernicus—for example, the definition of the length of a day, and the estimate of the size of the moon. From all this it is clear that Peucer is altogether at one with his teacher Erasmus Reinhold on the subject of astronomical hypotheses.

Yet there are certain texts which seem to run counter to this conclusion.

Kaspar Peucer wrote not only on the *Elements of the Doctrine of the Celestial Circles* but also composed a little book *On the Size of the Earth*.[16] In this work[17] Peucer writes as follows concerning the hypotheses governing the measurement of the earth:

Before taking up the questions that we propose to discuss, we must establish some hypotheses. We must let it be known that these hypotheses are not false, that they have not arbitrarily been devised for the purpose for which we intend to use them—they agree with the facts. They were originally discovered under the guidance and instruction of experience. Later they were secured and proved by demonstrations. It is of the greatest importance that the truth and certainty of these hypotheses be established with certainty and in detail, for if they are doubtful, ambiguous, or uncertain, the truth of everything that is constructed upon these foundations totters, crumbles, or is in danger of so doing. . . .

We must, then, from the outset, regard as true, fixed, and certain the two hypotheses which follow:

In the first place, the earth with the waters that surround and pervade it forms a single globe.

16. Peucer *De dimensione Terrae et geometrice numerandis locorum particularium intervallis ex doctrina triangulorum sphaericorum et canone subtensarum liber, denuo editus, sed auctius multo et correctius quam antea. Descriptio locorum Terrae Sanctae Exactissima, autore quodam Brocardo Monocho. Aliquot insignium locorum Terrae Sanctae explicatio et historiae per Philippum Melanthonem.* (Wittenberg, 1553.)

17. Ibid. "De hypothesibus, quas ut exploratas et demonstratas sequenti doctrinae praemittimus," pp. 17–23.

In the second place, the height of the highest mountains is insignificant in comparison to the dimensions of the globe.

Does not this language testify to the most intransigent realism with respect to hypotheses? Do we not here have an explicit rejoinder to the statements Osiander inserted into the preface to the *Six Books on the Revolutions*? Even though, in the dedicatory letter which introduces his book, Peucer referred to Erasmus Reinhold in eulogistic terms, he here seems clearly to diverge from his teacher's ideas, ideas which Ariel Bicard had adopted and which were seemingly accepted by Peucer himself in his *Elements of the Doctrine of the Celestial Circles.* The dedicatory epistle was addressed to Rheticus' son; was it, perhaps, to please *him* that Peucer here seems to outdo the author of the *Narratio prima* in realism?

The apparent contradiction between the statements successively made in Peucer's two works vanishes once one takes into account opinions that were very nearly universally accepted in the middle of the sixteenth century.

To begin with, there is a point that must certainly be granted Erasmus Reinhold's disciple: the two hypotheses on which he bases his geography are by no means fictions dreamed up for the sole purpose of saving the phenomena. They are propositions which claim to agree squarely with concrete reality: they are given out as *true.* Peucer was therefore not asking too much of the basic assumptions of geography.

But how can he, without risk of inconsistency, drop the requirement of truth when dealing with astronomical hypotheses? Let us recall the principles that Proclus, Maimonides, and Lefèvre d'Etaples had formulated so clearly, which, in a more or less explicit way, controlled the opinions of the Wittenberg school: The nature of the sublunary bodies lies within the mind's grasp. The physics of these bodies we can base upon propositions that are true and that conform to reality. But the nature of the celestial essence passes our understanding. We are unable, therefore, to deduce the movements of the stars from principles that are certain. We can do no more than establish astronomy upon fictive hypotheses whose sole object is to save the phenomena.

The view of hypotheses professed by the Wittenberg astronomers in the middle of the sixteenth century appears to us quite homogeneous: they all supported the doctrine Osiander formulated in his celebrated preface. The testimony of Melanchthon will show us presently that at this university the theologians thought exactly as the astronomers.

Nor were these ideas the exclusive property of the Wittenberg school; we find them at Nuremberg, with Schreckenfuchs, at Basle, with Wursteisen.

Thus, glancing through the voluminous *Commentary on Georg Purbach's New Theories of the Planets* brought out by Erasmus Oswald Schreckenfuchs in 1556,[18] we find that, when speaking of astronomical hypotheses Schreckenfuchs sounds pretty much like Proclus or Simplicius.

He lays down the principle that the movements of the heavenly bodies must be reduced to circular and uniform motions.[19] Guided by this principle,

> the ancients, wanting to save the appearances of the wandering stars, ascribed several movements to each one of them. Every such movement, considered in isolation, is uniform and always going in the same direction. But by compounding all these movements, a diversified movement is obtained. . . . From this it is clear that the one and only object of the doctrine of the *Theories* is to save the appearances of the wandering stars and to eliminate all irregularities in their movements.

Later on,[20] Schreckenfuchs, taking his inspiration from Melanchthon and Reinhold, notes that Purbach's order of demonstration proceeds according to τὸ ὅτι whereas Ptolemy arranged his demonstrations according to διότι:

> Ptolemy, having, through repeated observations, laid hold of the *causes* of the irregularities presented by the movements of the several wandering stars, applied all his intelligence to thinking up an ingenious arrangement of orbits that would, as they say, "save" this diversity. Purbach mastered this ordering of the orbits, as well as everything necessary for the mathematical study of the secondary mobiles, and, setting aside the geometrical demonstrations, he explained this branch of studies learnedly and clearly through the method called τὸ ὅτι to students of this marvelous science.

From what Schreckenfuchs here says about astronomical hypotheses one would conclude that his opinions differ very little from those so

18. Erasmus Oswald Schreckenfuchs *Commentaria in novas theoricas planetarum Georgii Purbachii* (Basel: per Henrichum Petri, 1556).

19. Ibid., p. 3.

20. Ibid., p. 4; cf. "Praefatio," toward the end.

clearly expressed by Osiander; this conclusion is strikingly confirmed when one watches how Schreckenfuchs employs these hypotheses.

Let us look at the interesting preamble to the third book,[21] which is devoted to the motion of the eighth sphere or, in more modern terms, to the precession of the equinoxes. First, Schreckenfuchs calls to mind the theories proposed by Ptolemy, Thabit ibn-Qurra, and the *Alphonsine Tables*.[22] He then continues as follows:

Finally, there came Nicholas Copernicus, that miracle of nature, and John Werner of Nuremberg. I shall not say here which of these two astronomers outranked the other in the minute study of the positions and motion of the eighth sphere. But I would openly declare that, whichever of these two we take as model, we will unquestionably make headway faster toward truth by imitating him than by imitating any other of the opinions we have just now passed in review.

Schreckenfuchs, we see, offers John Werner's theory of the equinoxes and Copernicus' theory as equally plausible, equally an advance over the theories of the ancients. Now John Werner's theory,[23] of which he gives a summary, is merely a modified version of the Alphonsine system: it keeps the earth immobile and at the center of the world. By contrast,

Copernicus, who inverted all the movements studied in astronomy, did the same to the eighth sphere: He regarded it as fixed and immobile; the true and the mean equator he imagines as below this sphere and as moving from the first star of Aries in a direction opposite to that of the constellations.

Clearly, the physical reality of astronomical hypotheses is of small concern to Schreckenfuchs. He does not care whether one assumes the earth immobile or sets it in motion, provided he is furnished with kinetic combinations that can save the displacements of the set of fixed stars exactly. In practice the commentator of Purbach adheres to the principles of Osiander.

Schreckenfuchs's teaching at the University of Nuremberg, like Reinhold's at Wittenberg, formed a group of disciples who did not ascribe any

21. Ibid., pp. 388–89.

22. These tables, prepared in the thirteenth century at the instigation of Alfonso X of Castile (whence their name), superseded earlier Muslim tables and served the European astronomers until they were in turn displaced by Reinhold's *Prutenicae tabulae* and others.—TRANSLATOR.

23. John Werner *Tractatus de motu octavae sphaerae et summaria enarratio theorica motus octavae sphaerae* (Nuremberg, 1522).

reality to astronomical hypotheses and required simply that they furnish correct astronomical tables. Christian Wursteisen (Vurstisius), who taught at Basle, was one of these disciples.

Wursteisen's *Questions concerning Georg Purbach's Theories of the Planets*[24] opens with an extremely interesting *Praefatio isagogica.* He there cites the suggestion of Pontano of which we spoke earlier, and he adopts the doctrine it entails as his own. He also mentions Proclus' *Hypotyposes.* No doubt it was Proclus who inspired him to reflections like the following:

Does each of the celestial spheres have as many orbits as astronomers assign? No one has ever been able to decide. The human mind merely conjectures that a given arrangement agrees with the natural effects and with observation. God alone knows the true causes, the order and arrangement of his noble and marvelous work. He offers us this work for contemplation, but of the knowledge he has of it he has sent us only a few rays. The heavens are spread out over the humble abode the earth furnishes to mortals. We do not dwell in the heavens. We can neither see them face to face nor touch them with our hands, and no one has come down to us from there to tell us what he has seen. . . . Concerning these objects, then, which do not fall under our senses, we shall hold that we have pushed our demonstrations far enough when we have reduced them to possible causes, that is to say, to causes from which nothing absurd follows.

The last sentence is borrowed from Aristotle's *Meteorology:* It makes more stringent demands on astronomical hypotheses than did Reinhold, Bicard, and Schreckenfuchs; it wants them to be at least *possible*, nothing *absurd* should follow from them. We shall soon see that this requirement would be taken to rule out the adoption of Copernicus' system.

Wursteisen himself does not seem inclined to derive justification for so intransigeant a position from the principles he laid down. In the matter of alternative hypotheses he seems rather to share the broad eclecticism of his teacher Schreckenfuchs—as the concluding sentences of his book tend to show. He has just explained the theory of the movement of the sphere of the fixed stars expounded in the *Alphonsine Tables* and by Georg Purbach; now he adds that this theory does not perfectly correspond with the phenomena:

24. Christian Wursteisen *Quaestiones novae in theoricas novas planetarum doctissimi mathematici Georgii Purbachii Germani, quae Astronomiae sacris initiatis prolixi commentarii vicem explere possint, una cum elegantibus figuris et isagogica praefatione* (Basel: ex officina Henricpetrina, 1568, 1573, 1596).

But I did not feel obliged to show this here, partly because this theory was so ingeniously conceived, and partly because it has provided great men with a serious ground for thinking up more solid doctrines; such were John Werner of Nuremberg and, above all, Nicholas Copernicus of Thorn. But this is not the place to discuss the subtle teachings concerning this portion of astronomy that they have left us.

Plainly, Wursteisen's attitude to Copernican hypotheses is like that of Osiander, Reinhold, and Schreckenfuchs.

What holds for the German Ptolemaists, namely, that they used astronomical hypotheses as Osiander wanted them to be used, holds all the more for the Italian Ptolemaists who came after Copernicus. Alessandro Piccolomini, for example, takes account of the principles enunciated in the famous preface; in fact, he formulates them in almost the very same terms. We should not be surprised that the partisans of Ptolemy adhered to the doctrine of Osiander, the disciple of Copernicus. It was this doctrine which, throughout antiquity and the Middle Ages, had constantly been their defense against Peripatetic and Averroist attacks.

Alessandro Piccolomini, in his *Theories of the Planets*,[25] of which only the first part was published, considers,

by way of a digression, whether the assumption thought up by astronomers for the purpose of saving the appearances of the planets have their foundation in the truth of nature.

It is thought by some that when Ptolemy, the astronomers he followed, and his successors imagined epicycles and eccentrics on the vault of the celestial sphere, they did so to have it really believed that this is how the orbits are arranged in the sky.

Those who think along these lines are forced into disputes about the possibility or propriety of such assumptions.

I do not want to stop to argue whether these inventions are possible or impossible, whether they are friends or enemies of nature, whether they are abhorrent to it or not. The possibility or impossibility of these contrivances makes them conform neither more nor less to the astronomers' intentions. For their intention consists exclusively in finding a way by which it is possible to save the appearances of the planets, to calculate

25. Alessandro Piccolomini, *La prima parte delle theoriche overo speculationi de pianeti* (Venice: appresso Giordan Ziletti, al segno della Stella, 1563), "Per modo di digressione si discorre se le imaginationi fatte da gli Astrologi per salvar le apparentie dei Pianeti sono fondate nel vero della Natura," chap. 10, fols. 22–23. (The first edition of this work appeared in Venice in 1558.)

and estimate them, to predict them from one time to another. But I would be so bold as to say that these critics are very much mistaken if they think that Ptolemy and his successors constructed these images, inventions, or combinations in the firm belief that this is how things are in nature. No, for these astronomers it was amply sufficient that their constructs save the appearances, that they allow for the reckoning of the movements of the heavenly bodies, their arrangements, and their places. Whether or not things really are constructed as they envisage them—that question they leave to the philosophers of nature; themselves they do not trouble with it, so long as their assumptions manage to save the appearances.

They know that from false premises one may deduce a true conclusion. They know that different causes can produce identical effects:

We observe a host of planetary appearances in the sky. The causes from which they really proceed remain unknown to us. But it is enough for us to be sure that if our inventions were true, the appearances which would derive from them would be no different than those we observed in fact. This amply suffices for purposes of calculation, for prediction, for the knowledge we desire—the situations, positions, sizes, and movements of the planets.

When, therefore, astronomers frame their assumptions, they hardly bother with the question whether the things they have imagined are necessary, probable, or false. This is why we find that Ptolemy, seeking to save the solar appearances, asserts and demonstrates that this may just as well be accomplished by an eccentric as by an epicycle. Of these two ways . . . he chose the eccentric, but he leaves others free to choose either, since the same effect can be seen to follow from both. Ptolemy would not have used this kind of language if he had thought that, in order for us to be able to deduce and conclude these appearances, the means he imagined must be true things of nature and the orbits arranged in the sky just as he distributed them.

Lucretius, Piccolomini adds, proceeded in the same way as Ptolemy when he studied the movements of the heavens:

He is satisfied to assign certain probable reasons, that is, reasons such that, if we suppose that they are true, the effects being considered follow necessarily. An effect cannot, of course, have more than one proper, true, and necessary cause. However, as I said earlier, the same effect may follow from several different causes, not merely probably but even necessarily. Not, admittedly, from the essential nature of these causes, but as a necessitated result and a logical consequence of the assumptions made. . . . This is what, by way of digression, I wanted to say against those who are in the habit of finding fault with good astronomers without knowing their intentions.

Osiander's doctrine in the preface to the *De revolutionibus* could hardly be expressed more clearly than it is here by Piccolomini. Moreover, Piccolomini obviously set great store by it, for before presenting it in his *Theories of the Planets* he taught the same doctrine somewhat more summarily but equally clearly and in almost the same terms in his *Natural Philosophy*.[26]

Andreas Cesalpinus, in his *Peripatetic Questions*, shows that he is a partisan of the Ptolemaic system. However, in one important point he proposes to make a modification of this system, which brings it closer to the systems of Copernicus and Tycho Brahe. He is aware that no combination of orbits allowed Ptolemy's successors to provide a satisfactory representation of the movements of Venus and Mercury. Cesalpinus wants, then, with respect to these two planets, to return to the ancient hypothesis of Heraclides Ponticus, Adrastus of Aphrodisias, and Theon of Smyrna: Venus and Mercury should be made to revolve around the sun. He adds:

> We shall not demonstrate in the present writing that the results which others have obtained by other means follow from this theory of the circles and movements of these stars as well. To do so would carry us beyond the confines of the terrain we here intend to cover. We do not thereby maintain that the statements of the astronomers are untrue. They consider natural bodies not in so far as they are natural but in the mathematical manner. For them it is, therefore, sufficient not to be mistaken as regards the calculations and predictions of movements (*idcirco satis est ipsis circa motuum numeros et supputationes non mentiri*). It is for the physicist to pursue these investigations according to the method of physics. Now the method of physics consists in this, that everything that occurs among the heavenly bodies be brought about in the same way and the means employed be as few as possible (*per pauciora magis quam per plura*).[27]

Cesalpinus does not restrict the astronomer's liberty of choice more narrowly than did Osiander. And even the physicist is confined only by

26. Piccolomini, *La seconda parte de la Filosofia naturale* (Venice: appresso Vincenzo Valgrisio, alla Bottega d'Erasmo, 1554), bk. 4, chap. 5, pp. 381–84.

27. Andreas Cesalpinus, *Peripateticarum quaestionum libri quinque* (Venice: apud Iuntas, 1571). lib. 3, quaest. 4, "Planetas in circulis, non in sphaeris moveri," fol. 57 (verso). We were unable to consult the first edition of this work, published in Florence in 1569.

principles which Ptolemy would undoubtedly have accepted: Posit analogous hypotheses in analogous cases! Give preference to the simpler hypotheses!

Francesco Giuntini, like Reinhold, was steeped in judicial astrology.[28] He computed tables which were offered as achieving a still greater precision than the *Prutenicae tabulae.* What value did he assign to astronomical hypotheses and, in particular, to the Ptolemaic hypotheses which he was constantly using? We must turn to his commentaries on John of Sacro-Bosco's *Treatise on the Sphere* to find out what his opinion on this subject was.

This investigation requires, however, some discernment. The art of commenting on John of Sacro-Bosco's *Treatise on the Sphere* frequently consists, for the somewhat unscrupulous Giuntini, in simply reproducing long passages borrowed from various astronomical writings. The only change he brings to the original text is the suppression of the author's name. Thus the two unequally developed commentaries that he wrote at different times contain whole pages culled from the *Quaestiones in libros de caelo et mundo* written by Albert of Saxony in the fourteenth century.

Giuntini's *Sphaera emendata,* which went into numerous editions after 1564,[29] contains a brief disquisition on astronomical systems.[30] But we must guard against seeing in this account a statement of Giuntini's own opinion, since it is merely an excerpt from a commentary on the *Treatise on the Sphere* which the Spaniard Pedro Sanchez Cirvelo of Daroca pub-

28. I.e., "The science by which man may know what will come to pass in the world or in this or that city or kingdom and what will happen to a particular individual all the days of his life." The definition quoted is Maimonides', who explains that "every one of those things concerning judicial astrology that [its adherents] maintain—namely, that something will happen one way and not another, and that the constellations under which one is born will draw him on so that he will be of such and such a kind and so that something will happen to him one way and not another—all those assertions are far from being scientific; they are stupidity. . . . Never did one of those genuinely wise men of the nations busy himself with this matter or write on it." The *Letter on Astrology* which we are citing is now readily available in Ralph Lerner and Muhsin Mahdi, *Medieval Political Philosophy: A Sourcebook* (New York: Free Press of Glencoe, 1963).—TRANSLATOR.

29. The edition we are using is the following: *Sphaera Joannis de Sacro Bosco emendata, cum . . . familiarissimis scholiis, nunc recenter compertis et collectis a Francisco Junctino Florentino sacrae Theologiae Doctore. Inserta etiam sunt Ellae Vineti Santonis egregia scholia in eandem Sphaeram.* (Lyons: apud haeredes Iacobi Iunctae, 1567).

30. Ibid., pp. 103–05.

lished in Paris in 1498 along with Pierre d'Ailly's *Fourteen Questions*[31] on the same book.

In 1577–78 a much more extensive though frequently equally derivative commentary of Giuntini's was printed in Lyon.[32] Here Giuntini does speak for himself on the subject of astronomical hypotheses, and he formulates his ideas quite clearly:[33]

It is not possible to demonstrate that the movements which appear in the heavens can be saved, except by using eccentrics and epicycles arranged as the astronomers assume them arranged.

Nevertheless, eccentric movements necessarily exist in the heavens.

Furthermore, until now no one has found a more reasonable method of giving the rule of each motion than that which uses eccentrics and epicycles.

In support of the first proposition, Giuntini appeals to the passages in Aquinas which we cited earlier. Aquinas' conclusion, namely, "hoc non est demonstratum, sed suppositio quaedam" (this has not been demonstrated but is merely an assumption), he adopts as his own.

On the other hand, that the planets do not have constant apparent diameters, from which it follows that they are not always at the same distance from the earth—this is *not* a mere assumption. It must, accordingly, be admitted that some celestial revolutions do not have the earth as their center:

And this second proposition does not contradict the first. For we did not say that it is impossible to prove the existence of eccentric movements.

31. Pierre d'Ailly *Uberrimum sphere mundi comentum intersertis etiam questionibus*. Colophon: Et sic est finis hujus egregii tractatus de sphera mundi Johannis de Sacro Bosco Anglici et doctoris Parisiensis. Una cum textualibus optimisque additionibus ac uberrimo commentario Petri Cirveli Darocensis ex ea parte Tarraconensis Hispanie quam Aragoniam et Celtiberiam dicunt oriundi. Atque insertis persubtilibus questionibus reverendissimi Domini Cardinalis Petri de Aliaco ingeniosissimi doctoris quoque Parisiensis. Impressum est hoc opusculum anno Dominice Nativitatis 1498 in mense februarii Parisius in campo Gallardo oppera atque impensis magistri Guidonis mercatoris. Cap. 4.

32. Giuntini *Sacrae Theologiae doctoris, Commentaria in Sphaeram Ioannis de Sacro Bosco accuratissima* (Lyons: apud Philippum Tinghium, 1578). This part contains the commentary on chapters 1 and 2 of Sacro-Bosco's *Sphaera*. Giuntini *Sacrae Theologiae doctoris. Commentaria in tertium et quartum capitulum Sphaerae Io. de Sacro Bosco* (Lyons: apud Philippum Tinghium, 1577). In this second part of his commentary, Giuntini repeats the plagiarism from Cirvelo (see pp. 304–4).

33. Ibid., commentaries in chap 4, pp. 330–43.

Rather, we said that the necessity of arranging them in the manner of Hipparchus, Ptolemy, and the modern astronomers cannot be demonstrated.

Giuntini goes on to prove that this is so by imagining kinetic combinations different from the ones Ptolemy proposed yet just as capable of saving the irregularities of the planetary movements.

All the same, to prove his third proposition, namely, that the system of eccentrics and epicycles of the *Almagest* is more reasonable than any other system, he makes use of Ptolemy's arguments.

Giuntini obviously agrees with what Piccolomini (whose book he mentions with approval) had said about astronomical hypotheses. Furthermore, at the beginning of his commentary,[34] he writes as follows:

Astronomy is divided into five parts.

The first part considers the movements, situations, and shapes of the heavenly bodies in general. This is the part with which the Philosopher dealt in the book *On the Heavens.* We must not, however, call it "astronomy," for it considers all these things not in terms of mathematical arguments but in terms of physical arguments.

The second part considers the movements, shapes, and situations of the heavenly bodies in general through mathematical arguments. This is what the author explains in the present treatise. In comparison to the other parts it is general.

The third part descends in particular to the movements of the planets and the revolutions of the heavenly bodies. This is the part Ptolemy dealt with in the *Almagest.*

The fourth part descends especially to the conjunctions, oppositions, and the aspects of the planets in their relation to one another. Ptolemy speaks of these things too in the *Almagest.* Subordinated to this part there are certain special studies, among them the construction of tables like the *Alphonsine Tables*, the *Tabulae prutenicae*, and our own, which are called *Tabulae resolutae astronomicae.*

The fifth part is judicial astrology.

The author of the *Tabulae resolutae astronomicae* clearly has the same attitude to astronomical hypotheses as the author of the *Tabulae prutenicae.*

Giovanni Battista Benedetti too was keenly interested in astronomical tables. His correspondence is full of comments about the *Tabulae prutenicae* and Giuntini's tables.[35] These comments induce him to refer fre-

34. Ibid., commentaries in chap. 1, p. 10.

35. Giovanni Battista Benedetti *Diversarum speculationum mathematicarum et physicarum liber* (Turin: apud haeredem Nicolai Bevilaque, 1585).

quently to the *De revolutionibus*.[36] Being an excellent geometer, he speaks with obvious admiration of the kinetic combinations Copernicus had proposed to save the celestial phenomena. But the Copernican *hypotheses* seem to concern him as little as his correspondents. Only once does he refer to these hypotheses,[37] not to adopt them, nor to reject them, but merely to recall that, for the Copernicans, the earth is reduced to the part of being the center of the lunar epicycle. Who knows, he adds, whether a body like the earth may not be at the center of every planetary epicycle?

The state of mind of the majority of astronomers during the twenty or thirty years following the publication of Copernicus' book seems quite clear: Copernicus' work quickly won their attention because it seemed eminently suited to the construction of exact astronomical tables and because the kinetic combinations of Copernicus seemed preferable to Ptolemy's. As for the hypotheses from which Copernicus had deduced his kinetic combinations and the question whether they are true, probable, or purely fictive—these matters they left to the physicists; it is the business of the philosopher of nature to settle such questions. They treated these hypotheses as Osiander had suggested they should, not because the anonymous preface had in any way imposed this attitude on them, but because this had for long been their customary attitude. From Greek antiquity through the entire Middle Ages and until the beginning of the Renaissance it was this attitude that had enabled the partisans of the Ptolemaic system to make advances in astronomy in spite of the Peripatetics and Averroists; they simply disregarded the latter's repeated and always futile efforts to restore the system of homocentric spheres. The astronomers who immediately followed Copernicus treated hypotheses in the manner of the Parisian and Viennese scientists of the fourteenth and fifteenth centuries; Schreckenfuchs and Reinhold continued the tradition of men like Purbach and Regiomontanus. This is why we find astronomers who use the geometric constructions of the *De revolutionibus* defending exactly the same views about the nature of astronomical assumptions as do those who continue to adhere to the methods of calculation of the *Almagest*.

And during this period, the theologians too shared this view. In this connection it is extraordinarily interesting to study Melanchthon's state of

36. Ibid., pp. 215, 216, 235, 241–43, 260, 261, 315.
37. Ibid., p. 255.

mind. Melanchthon taught at Wittenberg together with Reinhold. Reinhold's first book had a preface by him.

It was Luther who first declared war on the hypotheses of Copernicus—in the name of Scripture. Melanchthon, his faithful disciple, could not but follow him.

In 1549 Melanchthon published the lectures on physics he had been delivering at Wittenberg.[38] This is what he has to say about the hypothesis of the earth's movement:

Some have claimed that the earth moves. They assert that the eighth sphere and the sun remain immobile whereas they assign movement to the other spheres and count the earth among the stars. There is a book by Archimedes, called *De numeratione arenae,* in which the author reports that Aristarchus of Samos defended this paradox: the sun stays fixed and the earth turns around the sun.

Clever scientists take pleasure in debating a host of questions which give scope to their ingenuity. But young people should realize that these scientists have no intention of asserting such things. Let young people's primary allegiance be to opinions that have the benefit of the common assent of competent people, opinions that are not in the least absurd. They will then understand that God has revealed the truth; they should accept it with respect and acquiesce in it. [39]

Melanchthon, accordingly, tries to prove the earth's fixity not only by using the classical arguments of Peripatetic physics but also, and chiefly, by means of texts taken from Holy Scripture—the very arguments and texts which would, some eighty years later, be cited against Galileo.

The same Melanchthon who, in the name of physics and theology, so explicitly condemns the hypotheses of Copernicus has this to say about the moon:

I shall follow the customary method which comes down to us from Ptolemy and which most astronomers have followed up to now. Although the combination of lunar orbits recently thought up by Copernicus is extremely well adapted (*admodum concinna*), we shall nevertheless retain the Ptolemaic in order somehow to initiate students into the doctrine that is commonly accepted at the schools.[40]

How can Melanchthon, without flagrant contradiction, say that the Copernican hypotheses are contrary to physics and theology and still ad-

38. Melanchthon *Initia doctrinae physicae dictata in Academia Vuitebergensi*, 2d. edition (Wittenberg: Johannes Lufft, 1550). We were unable to consult the first edition of this work, published in 1549.

39. Ibid., bk. 1, cap., "Quis est motus mundi?" fols. 39–42.

40. Ibid., cap., "De Luna," fol. 63 (recto).

mire the theory of the moon which is deduced from these hypotheses? The reason is not far to seek. According to Melanchthon, Copernicus framed his hypotheses solely with a view to saving the phenomena. He is convinced that neither Copernicus nor his disciples meant to offer their assumptions as realities: "*Sciant juvenes non velle eos talia asseverare.*"

And it was quite natural for Melanchthon to attribute this sort of attitude to the Copernicans, since this was how he himself treated the Ptolemaic hypotheses, which he by far preferred.

Thus, writing about the sun's movement,[41] he says:

In order that we might somehow understand what this proper movement of the sun is, certain very learned geometers have manufactured kinds of automats. They stacked a certain number of spheres one inside the other and the planets were, so to say, lodged within these. It is even said that Archimedes constructed such αὐτόματα of the celestial movements, that is, orreries which represent these movements to the eyes. . . .

This is the place to censure the perversity and quarrelsome disposition of Averroes and many other philosophers. They make fun of this doctrine which is put together with so much art because we cannot say that such mechanisms really exist in the heavens.

If only Averroes and the others would stop bringing confusion into established science. Why do they not show us laws of the celestial movements which are better adapted and through which we might set up exact computations? Since Averroes' arguments are extremey crude (*prorsus* βάναυσα, we need not repeat them here. Besides, geometers themselves never meant to claim that such models exist in the heavens. They only want to give an exact amount of their movements.

Somewhat later Melanchthon reiterates this position.[42] Thus, immediately after the passage where he explains that he will deal with the moon's movements according to the method of Ptolemy, despite the accuracy of the Copernican theory, he adds:

In this connection it is proper to remind the listener that when geometers got the idea of constructing such spheres and such an epicycle, it was to make the laws of their movements and periods visible, and not at all because such a mechanism exists in the sky, although it is a matter of common parlance to say that there are certain spheres up there.

Since the only function of astronomical hypotheses, according to Melanchthon, is to represent the celestial phenomena and to facilitate their exact computation, the hypotheses themselves having no reality

41. Ibid., cap., "De Sole," fols. 52 (verso), 53 (recto).
42. Ibid., cap., "De Luna," fol. 63 (recto).

whatever, we should not be surprised when he says that the theories of Copernicus are very accurate, while at the same time rejecting them—specifically, the hypothesis of the movement of the earth—in the name of physics and Scripture.

We have found no text which allows us to learn how Catholic theologians contemporaneous with Melanchthon viewed astronomical hypotheses. But there is one highly significant fact which suggests that in general they agreed with Melanchthon on this subject: the computations that enabled Gregory XIII in 1582 to complete the reform of the calendar were based on the *Tabulae prutenicae*.[43] Certainly, in employing these tables, constructed by means of the theories of Copernicus, the Pope in no way intended to subscribe to the hypothesis of the earth's motion. He too looked upon astronomical hypotheses as contrivances ordered exclusively to the saving of appearances.

As time goes on, however, the hostility of theologians and philosophers toward the Copernican hypotheses increased. Like Melanchthon, they considered these hypotheses philosophically false and theologically heretical but, less tolerant than Melanchthon, they would not brook their use even in astronomy. Even the acclamation of Copernicus' astronomical genius offended them. In 1569 Schreckenfuchs writes:

> All kinds of debate can be stirred up on the subject of the earth's motion. We shall find such discussions in the book by Nicolaus Copernicus, a man of incomparable genius. I would have every right to call him the world's miracle were I not fearful of thereby offending certain men who, however correctly, hold excessively to judgments handed down by the ancient philosophers.[44]

At about the same time we hear Peucer, a pupil of Reinhold and Melanchthon, raising his voice against the use of Copernicus' hypotheses in astronomy while accepting his procedures of calculation:

> The absurdity, absolutely foreign to the truth, of these hypotheses of Copernicus is shocking.[45]

Elsewhere in the same book—*Hypotheses astronomicae seu theoricae planetarum*, published in 1571—he writes:

43. August Heller, *Geschichte der Physik von Aristoteles bis auf die neueste Zeit* (Stuttgart, 1882), vol. 1, p. 270.

44. Schreckenfuchs *Commentaria in sphaeram Ioannis de Sacrobusto,* entire (Basel: ex officina Henricpetrina, September, 1569), p. 36.

45. Cited by Leopold Prowe, *Nicolaus Copernicus*, vol. 1, pt. 2, p. 281.

I have set my hypotheses in agreement with Copernicus' observations and tables. As for the Copernican hypotheses themselves, I am of the opinion that they ought under no circumstance to be introduced into the schools.

Evidently, the attitude of those interested in astronomical questions was changing. Any hypothesis capable of saving the phenomena, even one that was, from the philosopher's point of view, neither true nor probable, was considered useful by Gemma Frisius, Osiander, and those who held with them; but henceforth, before a hypothesis could be employed in astronomy, it would be required to be—either certainly or more or less probably—in accord with the nature of things. From now on astronomy was to be subject to philosophy and theology.

7
From the Gregorian Reform of the Calendar to the Condemnation of Galileo

Astronomical hypotheses are simply devices for saving the phenomena; provided they serve this end, they need not be true nor even likely.

From the time of the publication of Copernicus' book with the preface by Osiander up to the time of the Gregorian reform of the calendar, this was, it seems, the generally accepted opinion of astronomers and theologians. During the half century that stretches between the reform of the calendar and the condemnation of Galileo, however, this conception of astronomical hypotheses becomes relegated to oblivion, or rather, it is furiously attacked in the name of the prevailing realism. The new realism insisted on finding declarations concerning the nature of things in astronomical hypotheses; it required, therefore, that they be in harmony with the teachings of physics and with scriptural texts.

The learned Jesuit Christopher Clavius of Bamberg wrote a lengthy commentary on the *Sphaera* of John of Sacro-Bosco. The first two editions of this book, printed in Rome in 1570 and 1575, did not go into the subject of astronomical hypotheses. But in 1581 Clavius prepared a third edition, "*multis ac variis locis locupletata.*"[1] On the *verso* of the title page he enumerates the additions; amongst them there is a "*disputatio perutilis de orbibus eccentricis et epicyclis contra nonnulos philosophos.*" That *disputatio*, entitled "*Eccentrici et epicycli quibus φαινομένοις ab astronomis inventi sunt in coelo,*" is quite lengthy; it takes up twenty-

1. Christopher Clavius *In Sphaeram Ioannis de Sacro Bosco commentarius nunc iterum ab ipso Auctore recognitus, et multis ac variis locis locupletatus.* Permissu superiorum (Rome: ex officina Dominici Basae, 1581).

seven pages of very fine print.[2] What is more, it is extremely interesting, because not only the Ptolemaic system, but also the Copernican hypotheses are examined. And Clavius was an admirer of Copernicus' work. In treating of astronomical inventors, he mentions it several times by name; both the *De revolutionibus orbium coelestium* and the *Tabulae prutenicae* are mentioned; he goes so far as to call Copernicus "that most excellent geometer who, in our time, has put astronomy on its feet again and who will, in recognition thereof, be celebrated and admired by all posterity as Ptolemy's equal." These sentiments give an especial weight to Clavius' critique of the Copernican hypotheses.

One additional circumstance enhances the importance of these criticisms: As a member of the Society of Jesus, Clavius was a part, as he tells us,[3] of the commission set up by Gregory XIII to prepare the reform of the calendar. He may, therefore, be considered an authoritative interpreter of the intellectual tendencies that prevailed in Rome at this time.

Clavius explains,[4] only to reject it, the opinion that turns the eccentrics and epicycles into fictions devised solely to save the appearances:

> Certain authors agree that all the φαινόμενα can be defended by assuming eccentric orbs and epicycles, but in their opinion it does not follow that these orbs really exist in nature; they are altogether fictive; there may in fact be some other more convenient method of defending all the appearances though it be as yet unknown to us. Besides [say they], it may very well happen that the true appearances can be defended by means of these orbs despite their being entirely fictive and no true causes of the appearances at all; for, as the dialectic of Aristotle shows, from the false one may infer the true.
>
> This reasoning receives additional confirmation from the following: In the work entitled *De revolutionibus orbium coelestium* Nicolaus Copernicus saves all the φαινόμενα in a different way. He assumes that the firmament is fixed and immobile; he further assumes that the sun, immobile, is at the center of the universe; as for the earth, he attributes a triple motion to it. The eccentrics and epicycles are, then, not necessary for saving the φαινόμενα of the wandering stars.

Clavius refuses to surrender to the force of these arguments. Of those who uphold them he says:

> If they have a more convenient method, why do they not show it to us? We would be satisfied with it and greatly beholden to them. What astronomers are after is to save all the celestial φαινόμενα in the most con-

2. Ibid., p. 416–42.
3. Ibid., p. 61.
4. Ibid., pp. 434–35.

venient manner, whether by the procedure of eccentrics and epicycles or by some other procedure. But since up to now no one has found a more convenient method than the one which saves the appearances by means of eccentrics and epicycles, it stands to reason that the celestial spheres have orbits of this kind.

If one should urge against Clavius that the reality of hypotheses cannot be proved from their agreement with phenomena so long as the impossibility of other hypotheses' saving these same appearances has not been established, Clavius would vigorously reject such an objection, saying that it would destroy the whole of physics, for this science is built entirely by proceeding from effects to causes. Sixty years earlier Luiz Coronel had, indeed, suggested just this, that physical theory should be assimilated to the doctrines of astronomy.

The fact that Copernicus had succeeded in saving the appearances by means of a system distinct from the Ptolemaic does, nevertheless, lead Clavius to attenuate his realist pronouncements, almost to reduce them to Giuntini's formulation:

That Copernicus should have succeeded in saving the *φαινόμενα* in a different way is not at all surprising. The motions of the eccentrics and epicycles taught him the times, the magnitudes, and the quality of appearances, future as well as past. Since he was exceedingly ingenious, he was able to conjure up a new method, in his opinion more convenient, of saving the appearances. . . . Just as, when we know a correct conclusion, we can construct a chain of syllogisms which derive that conclusion from false premises. But far from leading us to abandon eccentrics and epicycles, the doctrines of Copernicus would rather force us to assume them. Astronomers have imagined such orbs because the phenomena have taught them in a manner more than certain that the wandering stars do not always stay at the same distance from the earth. . . . All that can be concluded from Copernicus' assumption is that it is not absolutely certain that the eccentrics and epicycles are arranged as Ptolemy thought, since a large number of *φαινόμενα* can be defended by a different method. Now as regards this question, all we have tried to convince the reader of is that the wandering stars in their course do not always stay at one unvarying distance from the earth; so that there must be epicycles and eccentric orbits in the sky arranged as Ptolemy proposes or, at least, so that some cause must be placed there which, considered in terms of accounting for the effects, is equivalent to the eccentrics and epicycles.[5]

This conclusion repeats, practically word for word, the cautiously formulated proposition of Giuntini.

The Copernican system is just such a one as is here spoken of—it fur-

5. Ibid., p. 436–37.

nishes causes which, considered strictly in terms of accounting for astronomical phenomena, *are* equivalent to the eccentrics and epicycles. To conform to the rule he has just laid down, Clavius should have regarded the Copernican system as being as acceptable as the Ptolemaic:

If the Copernican assumption implied nothing false or absurd, one might, so long as it were a question of preserving the φαινόμενα, be in doubt whether it is better to adhere to the opinion of Ptolemy or to that of Copernicus. But the Copernican theory contains many absurd or erroneous assertions: it assumes that the earth is not at the center of the firmament; that it moves with a triple motion—a thing I find inconceivable, since, according to the philosophers, a single simple body has by rights a simple motion; [it further assumes] that the sun is at the center of the world and that it is bereft of any motion—all these things clash with the commonly accepted doctrine of philosophers and astronomers. Moreover, as we saw more fully in the first chapter,[6] these assertions seem to contradict what Holy Scripture in many places teaches us. This is why it seems to us that Ptolemy's opinion should be given preference over the opinion of Copernicus.

From these considerations there follows the following conclusion: It is probable that there are eccentrics and epicycles; it is just as probable that there are eight or ten heavens; for it was by means of the φαινομένοις that astronomers discovered this number of heavens and these orbs.

The position Clavius takes on the subject of astronomical hypotheses can, accordingly, be delimited in terms of the following propositions:

Astronomical hypotheses should save the phenomena as exactly and conveniently as possible, but this is not sufficient to render them acceptable.

One cannot make certainty a condition of acceptability; still, one should insist on probability.

To be probable, astronomical hypotheses must be compatible with the principles of physics and, besides, may not contradict either the teachings of the church or scriptural texts.

Thus, two conditions come to be imposed on any astronomical hypothesis that would make its entry into science:

It may not be *falsa in philosophia.*

It may not be *erronea in fide*, nor, a fortiori, *formaliter haeretica.*

6. In the first chapter, discussing the Copernican hypothesis of the earth's motion, Clavius, defending the immobility of our globe, says: "The Sacred Scriptures likewise support this opinion, for in many places they affirm that the earth is at rest whereas the sun and the other stars are in motion. (Favent huic quoque sententiae Sacrae Literae quae plurimis in locis Terram esse immobilem affirmant Solemque ac caetera astra moveri testantur.)" A list of the familiar relevant texts follows. Ibid., p. 193.

These are the very criteria by which the Inquisition would, in 1633, judge the two fundamental hypotheses of the Copernican system; it was because both seemed to the Holy Office *falsae in Philosophia* and the one *ad minus erronea in Fide*, the other *formaliter haeretica*, that Galileo would be prohibited from upholding them.

Three years before these two characteristics of any permissible astronomical hypotheses were suggested in the work that the Jesuit Christopher Clavius published in Rome, they were described and used at the other end of Europe by the Protestant Tycho Brahe.

Although Brahe's work on the comet of 1577[7] was not published until 1588,[8] the first eight chapters were completed by 1578. Now at the beginning of the eighth book Brahe explains, to justify submission of a new theory, why he believes he must reject both the system of Ptolemy and that of Copernicus.[9]

By assuming that the rotation of a planet's deferent is uniform, not around the center of that deferent, but around the center of the equant, Ptolemy had adopted "hypotheses that violate the first principles of the art." Brahe therefore took account of:

the innovation in the spirit of Aristarchus of Samos that was recently introduced by the great Copernicus. . . . This innovation expertly and completely circumvents all that is superfluous or discordant in the system of Ptolemy. On no point does it offend the principles of mathematics. Yet it ascribes to the earth, that hulking, lazy body, unfit for motion, a motion as quick as that of the ethereal torches, and a triple motion at that. By this it stands refuted, not only in he name of the principles of physics, but also in the name of the authority of Holy Scripture. For the latter, as we shall elsewhere show more fully, several times affirm the immobility of the earth. . . .

To me, therefore, both kinds of hypotheses (those of Ptolemy and those of Copernicus) seemed to involve serious difficulties. I meditated and searched myself deeply for some hypothesis that would be rigorously established in all respects—from the mathematical standpoint as well as

7. Cf. Houzeau and Lancaster, *Bibliographie générale de l'astronomie*, vol. 1, p. 596.

8. Tycho Brahe *De mundi aetherei recentioribus phaenomenis liber secundus, qui est de illustri stella caudata anno 1577 conspecta* (Uraniborg, 1588). Our citation of the work follows the text as reprinted in *Tychonis Brahe mathim: eminent: Dani Opera omnia sive Astronomiae instauratae progymnasta in duas partes distributa, quorum* (sic) *prima de restitutione motuum Solis et Lunae, stellarumque inerrantium tractat. Secunda autem de mundi aetherei recentioribus phaenomenis agit* (Frankfurt, impensis Ioannis Godofredi Schönwetteri, 1648).

9. Ibid., pt. 2, p. 95.

the physical, and that would not have to resort to subterfuge to avoid theological censure; [I sought,] in short a hypothesis fully adequate to the celestial phenomena.

The principles that Osiander had laid down in his famous preface now looked to Tycho Brahe like a mere subterfuge designed to evade theological censure. Astronomical hypotheses should not only save the phenomena; they should conform to the principles of Peripatetic philosophy and to Holy Writ as well; for they are not an expression of mere fictions, they describe realities. The hypotheses of Copernicus, however well adapted to the appearances, should be rejected because they cannot be brought into conformity with the nature of things. Tycho Brahe would say so again in the work that was, through Kepler's exertions, published one year after his death:

The arrangement which the great Copernicus attributed to the apparent rotations of the heavenly bodies is extremely ingenious and well adapted but it does not, in reality, correspond to the truth.[10]

Toward the end of the sixteenth and in the first years of the seventeenth century, Brahe's opinions on the nature of astronomical hypotheses spread in Germany.

We have before our eyes the manuscript of an unpublished little treatise on astronomy written on the model of Sacro-Bosco's *Sphaera;* George Horst of Torgau composed it in 1604, in Wittenberg.[11] Despite its elementary, textbook character, or rather because of it, this little work is singularly apposite for letting us know how astronomical hypotheses were viewed, in the early seventeenth century, at the celebrated Protestant university. It allows us to gauge how great a change of attitude had been produced in the fifty years since Melanchthon and Reinhold taught at that university. At the beginning of his little treatise, George Horst says:

Astronomy is the science of the motions to which the heavenly bodies are subject, either in relation to one another or in relation to the earth. It is called "science at its best" (*scientia a potiori*), for though it shows some of the things that are in the heavens only through sight (*κατ'ὄψιν*), it yet establishes most of its conclusions by means of apodictic principles

10. Brahe *Astronomiae instauratae progymnasta, quorum haec prima pars de restitutione motuum Solis et Lunae stellarumque inerrantium tractat* (Uraniborg, 1589; absoluta Prague, 1602), in Brahe *Opera omnia,* pt. 1, pt. 4.

11. George Horst *Tractatus in arithemeticam Logisticam Wittebergae privatim propositus* (1604); *Horst Introductio in Geometriam; Explicatio brevis ac perspicua doctrinae sphaericae in quatour libris distributa.*

and does this in a manner so certain and infallible that Pliny . . . rightly says: "It is shameful that there should be anyone who can bring himself to not believing in it."

The principles of astronomy are of two kinds—true principles and analogical principles. The former are arithmetic and geometry: by means of these sciences, as if by wings, we raise ourselves up to the sky and traverse it in flight in the company of the sun and the other stars. The latter are φαινόμενα and ὑποθέσεις: they are called analogical because they do not show that through which (*propter quid*) something exists or happens but only demonstrate *that* something happens. . . .

All things in the sky that present themselves to observation through sight are called φαινόμενα.

Hypotheses are assumptions made by scientists, assumptions by means of which they save and excuse the various φαινόμενα that are produced in the sky. Through them the man of science, who by nature desires to know the cause (τοῦ αἰτίου), as Aristotle says in the first book of the *Metaphysics*, comes to know the causes of these heavenly changes and to reveal them to others. Amongst the hypotheses we find the eccentric orbs, the epicycles, and other similar objects.

To these hypotheses, just as to the phenomena, George Horst ascribes absolute, apodictic certainty. To make sure that nothing throws this certainty in doubt, he takes great pains to enumerate, and to give a precise formulation of, all the hypotheses—concerning the sky, water, and the earth, and so forth—to which he subscribes. To each hypothesis the reasons that serve as its warranty are adjoined; these reasons are, almost always, arranged in two series: the author enumerates first those furnished by observation and Peripatetic physics, then those drawn from scriptural texts.

The immobility of the earth, for example, is confirmed by these two kinds of arguments, just as it was in the *Initia Physicae* of Melanchthon. But Melanchthon, in invoking these two kinds of proof in support of physical truth, left the astronomer free to save the phenomena by means of artificial hypotheses which are not in conformity with this truth. George Horst takes the hypotheses of astronomy for certain and infallible principles. That is why he tries to justify them by physical and theological argumentation.

The enemies of the Copernican system came to rely ever more heavily on this principle that astronomical hypotheses are an expression of physical reality. One might think that *their* attitude should have forced the Copernicans to take the opposite position, to maintain, with Osiander, that astronomical hypotheses are mere contrivances for saving the phenomena; by acknowledging that astronomical hypotheses should conform to the nature of things, the Copernicans imperilled their system. On the one hand, their assumptions contradicted precisely those principles of Peripatetic physics which were regarded as certain by the majority of philosophers, and they destroyed these principles without offering anything to take their place; the hypothesis of the earth's motion, for instance, was irreconcilable with scholastic teaching concerning the motion of projectiles, and no Copernican had tried to provide a new theory of this kind of motion. On the other hand, the motion of the earth and the immobility of the sun seemed explicitly denied by Holy Scripture, and this objection could not but appear to have great force to men who were for the most part sincere Christians, whether Catholic or Protestant.

Thus the Copernicans had every conceivable reason to incline toward the position recommended by the preface to the *De revolutionibus*. Yet the contrary position was the one they chose. With considerably more ardor than the Ptolemaists, they took it upon themselves to proclaim that astronomical hypotheses must be truths, and that the assumptions of Copernicus alone conform to reality.

Bruno is not just passionate when, in one of the earliest of his writings,[12] he combats Osiander's opinion; he is violently rude.

He reports that according to some:

> Copernicus did not really adopt the opinion that the earth is in motion, since this is a paradoxical and impossible assumption. Rather, he is supposed to have ascribed motion to the earth instead of to the eighth sphere solely with an eye to ease of calculation.

But, says Bruno:

> If Copernicus had affirmed the motion of the earth solely on this ground and not for any other reason, it would seem minor, even insignificant. But there can be no question that Copernicus believed in this motion, just as he affirmed it, and that he proved it with all the skill at his command.

Bruno thereupon speaks of

> a certain preliminary epistle affixed to Copernicus' book by I know not what ignorant and presumptious ass who wanted, it seems, to excuse

12. Michel di Castelnuovo *La cena de le ceneri. Descritta in cinque dialogi, per quattre interlocutori, contre considerationi, circa doi suggetti, all'unico refugio de le Muse* (1548). Reprinted in *Le opere italiane di Giordano Bruno* (Göttingen: Paolo de Lagarde, 1888), vol. 1, pp. 150–52.

the author; or rather, he wanted to make sure that even in this book other asses would find the lettuce and vegetarian fare he had left there so that they would not risk going off without breakfast.

After this courteous introduction, Bruno quotes from the preface, and goes on:

Behold the handsome doorman! Behold how good he is at opening the door to let you enter and participate in this most honorable science without which the arts of counting and of measuring, geometry and *perspectiva*, would be no more than a pastime for ingenious madmen. Marvel how faithfully he serves the master of the house!

Despite the poor taste of these sarcastic remarks, Bruno was right when he denounced the contradictions between Osiander's preface and Copernicus' letter to Pope Paul III. He was right when he claimed that Copernicus

took on the office, not only of the mathematician, who assumes the motion of the earth, but also that of the physicist, who demonstrates it.

Bruno's own realism is quite in the tradition of Copernicus and Rheticus.

Of that tradition, the most resolute and illustrious representative is, unquestionably, Kepler. Even in the preface to his first work, the *Mysterium cosmographicum*,[13] printed in 1596, he tells us that six years earlier, at Tübingen, when he was assistant to Michael Maestlin, he had already been captivated by the system of Copernicus:

From that time on, I resolved to attribute to the earth not only the motion of the first mobile, but also that of the sun. And whereas Copernicus does this for mathematical reasons, I attribute the sun's motion to the earth for physical or, if you will, metaphysical reasons.

Kepler was a Protestant, but deeply religious. He would not consider the Copernican hypotheses in conformity with reality if they were contradicted by Holy Scripture. Before he can advance on the terrain of metaphysics or physics, he must, therefore, traverse that of theology. At the beginning of Chapter I of the *Mysterium cosmographicum* he tells us that "in this discussion of nature we must, from the outset, take care not to say anything that is contrary to Holy Scripture."[14]

13. Kepler *Prodromus dissertationum cosmographicarum continens mysterium cosmographicum de admirabili proportione orbium coelestium deque causis coelorum numeri, magnitudinis, motuumque periodicorum genuinis et propriis, demonstratum per quinque regularia corpora geometrica* (Tübingen: excudebat Georgius Gruppenbachius, 1596) in Kepler *Opera* (Frisch ed., vol. 1, p. 106).

14. Ibid., p. 112.

Kepler here indicates the way the Copernicans will henceforth be obliged to follow. As realists they want their hypotheses to conform to the nature of things; as Christians they acknowledge the authority of the Holy Writ; they must, therefore, try to reconcile their astronomical doctrines with Scripture and are forced to set themselves up as theologians.

Had they conceived of astronomical hypotheses in the manner of Osiander, they could have avoided such constraint. But those who faithfully followed the suggestions of Copernicus and Rheticus could not endure the doctrine expounded in the famous preface. Says Kepler:[15]

> Certain individuals make much of an example drawn from an exceptional demonstration, namely, one in which a true conclusion is made to follow from false premises by rigorous syllogistic deduction. They claim, on the strength of this example, that the hypotheses entertained by Copernicus might be false and that nevertheless the true φαινόμενα could follow from them as from their proper principles. I have never been able to share this opinion. . . .
>
> All that Copernicus discovered a posteriori, all that he, by means of geometric axioms, demonstrated through sight, can, I do not hesitate to assert, be demonstrated a priori in a manner that would exclude all doubt and would even win the support of Aristotle, were he still alive.

As we saw earlier,[16] Ryemer Baer published his *De hypothesibus astronomicis* in 1597. In this work the doctrines which Osiander had expounded in the preface to the book *On the Revolutions* were again taken up. But to go by Kepler's analysis of Ursus' *De hypothesibus*,[17] Ursus disfigured the ideas of Copernicus' editor by extremely misleading exaggerations. One might, for example, read there[18] that "the hypotheses [of astronomy] are a fictive description of an imaginary form of the world system and not the real and true form of this system"—an idea which Lefèvre d'Etaples had developed magnificiently. But one would also read that "the hypotheses [of astronomy] would not be hypotheses if they were true," or that "the proper object of hypotheses is to let the true follow from the false." These assertions are a mere play on words. Even if the word "hypothesis" had in ordinary discourse acquired this sense of dubious assumption, philosophers and astronomers preserved its etymological meaning, namely, that of a basic proposition on which a theory rests.

To refute Ursus, Kepler composed, towards 1600 or 1601, a work

15. Ibid., p. 112–13.
16. See chap. 6, n. 2, above.
17. We were unable to consult the work ourselves.
18. Kepler, *Opera* (Frisch ed., vol. 1, p. 242).

which was never completed and was not published until recently.[19] This essay has already provided us with important historical information concerning the preface that opens the *De revolutionibus orbium coelestium*. We shall now quote from it to show what exactly Kepler's opinion as to the nature of astronomical hypotheses was:

In astronomy, as in every other science, the conclusions we teach the reader are offered him in all seriousness; mere plays upon words are excluded. We intend, therefore, to convince him of the truth of our conclusions. Now if a syllogism is to lead legitimately to a true conclusion, its premises—here the hypotheses—must be true. We do not, therefore, attain our end, which is to exhibit the truth to the reader, unless we set out from two hypotheses both of which are true so as to arrive, by the rules of the syllogism, at the conclusion. If error has entered, either into one of the two hypotheses that have been taken as premises or into both, it is quite possible that a correct conclusion follow, but as I said earlier, in Chapter I of my *Mysterium cosmographicum*, this would happen only by chance, and not always. . . .

There is a proverb that says: "Liars need a good memory." The same holds for false hypotheses that have accidentally led to a correct conclusion. In the course of demonstration, as they come to be applied to ever more varied cases, they will not always preserve that habit of furnishing true conclusions. They will surely end up by betraying themselves. . . .

Now none of the authors of hypotheses to whom we accord fame would wish to expose himself to the risk of erring in his conclusions. It follows that not one of them would want to adopt scientifically, among his hypotheses, a proposition infected with error. This is why one frequently finds them more solicitous of the hypotheses that are to be laid down than of what follows from the demonstrations, the conclusions. All the celebrated authors that have appeared to this day examine their hypotheses with the help of reasons furnished as much by geometry as by physics, and they want to make them agree as much with the one as with the other.[20]

But are there not *distinct* and yet *equivalent* hypotheses? Hypotheses that cannot simultaneously be true but that lead to identical conclusions? The theorem of Hipparchus, which allows the solar motion to be represented just as well by an eccentric as by an epicycle revolving on a circle concentric with the world, provides a classic example. Is this not proof positive that true conclusions can be deduced from a hypothesis though no astronomer could tell whether or not the hypothesis itself is true?

19. Kepler *Apologia Tychonis contra Nicolaum Raymarum Ursum* in Kepler *Opera* (Frisch ed., vol. 1, p. 215).

20. Ibid., p. 239.

In Kepler's opinion, this uncertainty is the portion of those astronomers who, in examining hypotheses, refuse to call on any except mathematical reasons; the simultaneous employment of reasons from geometry and reasons from physics will surely make it vanish:

He who weighs all things by this rule will, I have no doubt, never chance to encounter a single hypothesis—simple or complex—that will not, in the end, yield a particular conclusion separate and different from the conclusions that any other hypothesis might have provided. Though the conclusions of two hypotheses coincide in the domain of geometry, in the physical domain each will engender a special result. Scientists, however, do not always pay attention to these differences that become manifest only in the domain of physics; all too often they cramp their thought and will not let it go beyond the confines of geometry and astronomy. It is while staying within the limits of these two sciences that they discuss the question of equivalence of hypotheses. They abstract from the different consequences which, if the neighboring sciences were taken into consideration, would lessen or even eliminate that pretended equivalence.

According to Kepler, then, the equivalence of two distinct hypotheses can only be a partial equivalence. If certain conclusions be deducible from both of two irreconcilable hypotheses, it is not on account of their differences but in virtue of what they hold in common.

Here we reencounter the thoughts of Adrastus of Aphrodisias and Theon of Smyrna.

Kepler is not content to criticize the doctrine upheld by Osiander and Ursus. He means, further, to practice that realism whose principles he has laid down. To this realism the greatest memorial to his genius, the *Epitome Astronomiae Copernicanae*, bears witness.

It declares itself as soon as the book begins: "Astronomy," Kepler says, "is a portion of physics."[21] That this aphorism is far from innocuous is immediately apparent from what the author tells us *de causis hypothesium:*[22]

The third part of the astronomer's "baggage" is physics. It is not generally considered necessary for the astronomer; and yet the science of the astronomer has a great bearing on the object of this portion of philosophy, which, without the astronomer, could not reach completion. Astron-

21. Kepler *Epitome Astronomiae Copernicanae usitata forma quaestionum et responsionum conscripta, inque VIII libros digesta, quorum hi tres priores sunt de doctrina physica* (Lenz: excudebat Johannes Plancus, 1614) in Kepler *Opera* (Frisch ed., vol. 6, p. 119).
22. Ibid., p. 120–21.

omers should not, in fact, be given absolute license to feign anything whatever without sufficient reason. You ought to be able to provide probable reasons for the hypotheses you claim as the true causes of appearances. You ought, therefore, at the outset, to seek the foundations of your astronomy in a higher science, I mean, in physics or metaphysics. Then, sustained by the geometric, physical, or metaphysical arguments that your particular science has provided you with, you will not be prohibited from passing beyond the boundaries of that science and you may then discourse about the things that pertain to these higher doctrines.

In the course of the *Epitome*, Kepler takes every possible occasion to support his hypotheses with arguments drawn from physics and metaphysics. And what physics, what metaphysics! But this is hardly the place to tell what strange reveries, what childish fancies Kepler designated by these two words. We do not wish to investigate how Kepler in fact constructed his astronomy; for us it suffices to know how he wanted it to be constructed: He wanted, as we now know, the science of the celestial motions to rest on foundations guaranteed by physics and metaphysics; he further required that astronomical hypotheses be in harmony with Scripture.

But over and beyond this, a new ambition declares itself in Kepler's writings: Astronomy, once it is founded on hypotheses that are true, will, through its conclusions, be able to contribute to the advancement of physics and metaphysics, the very physics and metaphysics that initially supplied its principles.

To begin with, Galileo adopted the hypotheses of Ptolemy. In 1656, a little treatise on cosmography by the great Pisan geometer was printed in Rome.[23] It was included in the second volume of the Padua edition of Galileo's works that was published in 1744.[24] A short note by the editor indicates the existence of a manuscript copy of this same *opusculum*. According to this manuscript, Galileo wrote the work in 1606, to serve as a handbook to students at the University of Padua. Later editions of Galileo's works reproduce the little treatise.

Two years earlier, George Horst had brought out his *Expositio doctrinae sphaericae*, in Wittenberg. It is extremely interesting to compare Galileo's *opusculum* with the *Expositio* of George Horst. The dominant tendencies of the two authors are very much alike. Like Horst, Galileo

23. Galileo Galilei *Trattato della sfera o Cosmografia* (Rome, 1656).

24. *Opere di Galileo Galilei divise in quattro tomi, in questa nova edizione accresciute di molte cose inedite* (Padua, 1744), vol. 2, p. 514.

speaks first of the various factors that go into the makeup of astronomy. He singles out *phenomena*, then *hypotheses*. Like Horst, he offers a definition of hypotheses: "Certain assumptions bearing on the structure of the celestial orbs and such as to answer to the appearances," and continues:

> Since we are now dealing with the first principles of this science, we shall bypass the more difficult calculations and demonstrations and deal solely with hypotheses. We shall concentrate on confirming them and establishing them by means of the appearances.

What exactly does Galileo have in mind in speaking of the *confirmation* of hypotheses? Is it sufficient if they *save the appearances,* or must they be true or at least likely? Galileo's requirements are as stringent as Horst's: He too wants the foundations of astronomical theory to conform to reality. Like Horst, he claims to demonstrate their truth by means of the classical proofs of Scholastic physics. There is only one notable difference between Galileo's demonstrations and those of Horst: whenever he can, the Protestant professor at the University of Wittenberg supplements the justifying reasons drawn from the physics of Aristotle with the force of scriptural texts. The Catholic professor at the University of Padua never appeals to these texts.

When Galileo at last adopted the system of Copernicus, he did so in the same spirit that had inspired him while he held to the system of Ptolemy: the hypotheses of the new system were not to be mere contrivances for the calculation of astronomical tables but propositions that conform to the nature of things. He wanted them established on the ground of physics. One might go so far as to say that this *physical* confirmation of the Copernican hypotheses, is the center towards which all, even the most diverse, of Galileo's investigations tend. His observations as an astronomer and his theories as a student of mechanics converge toward this same end. Further, since he wanted the foundation for the Copernican theory to be truths, and since he did not think that a truth could contradict Scripture (whose divine inspiration he recognized), he was bound to attempt to reconcile his assertions with biblical texts. In time he too turned theologian, as is shown by his famous letter to Marie Christine of Lorraine.

In claiming that the hypotheses of astronomy express physical truths, in declaring that they do not seem to him to contradict Holy Writ, Galileo was, like Kepler, entirely in the tradition of Copernicus and Rheticus. He set himself against those who represented the tradition of Tycho

Brahe, the Protestant, and Christopher Clavius, the Catholic. What these had said around the year 1580, the theologians of the Holy Office solemnly proclaimed in 1616.

They seized on these two fundamental hypotheses of the Copernican system:

Sol est centrum mundi et omnino immobilis motu locali. Terra non est centrum mundi nec immobilis, sed secundum se totam movetur, etiam motu diurno.

They asked themselves whether or not these two propositions bear the two marks which Copernicans and Ptolemaists, with one accord, required of any admissible astronomical hypothesis: Are these propositions compatible with sound physics? Are they reconcilable with divinely inspired Scripture?

Now for the Inquisitors, sound physics was the physics of Aristotle and Averroes, which dictated an unequivocally negative reply to the first question: the two incriminated hypotheses were *stultae et absurdae in Philosophia.*

As for Scripture, the advisers to the Holy Office refused to accept any interpretation that did not have the authority of the church fathers on its side. Hence the answer to the second question was inescapable: the first proposition was *formaliter haeretica*, the second *ad minus in fide erronea.*

The two censured propositions bore neither of the two marks that were supposed to distinguish any admissible astronomical hypothesis; both must, therefore, be totally rejected; neither was to be used, even for the sole purpose of *saving the phenomena.* Thus Galileo was prohibited from teaching the doctrine of Copernicus *in any manner whatever.*

The condemnation carried through by the Holy Office resulted from the clash between two realist positions. This head-on collision might have been avoided, the debate between the Ptolemaists and the Copernicans might have been kept to the terrain of astronomy, if certain sagacious precepts concerning the nature of scientific theories and the hypotheses on which they rest had been heeded. These precepts, first formulated by Posidonius, Ptolemy, Proclus, and Simplicius, had, through an uninterrupted tradition, come down directly to Osiander, Reinhold, and Melanchthon. But now they seemed quite forgotten.

There were, however, voices of authority to call attention to them once again.

One of these was Cardinal Bellarmine's, the same who, in 1616, was to examine the Copernican writings of Galileo and Foscarini. As early as April 12, 1615, Bellarmine had written Foscarini a letter full of wisdom and prudence.[25] We quote from it below:

It seems to me that Your Reverence and Signor Galileo would act prudently by contenting yourselves with speaking *ex suppositione* and not absolutely, as I have always believed Copernicus to have spoken. To say that by assuming the earth in motion and the sun immobile one saves all the appearances better than the eccentrics and epicycles ever could is to speak well indeed. This holds no danger and it suffices for the mathematician. But to want to affirm that the sun really remains at rest at the world's center, that it turns only on itself without running from East to West, and that the earth is situated in the third heaven and turns very swiftly around the sun, that is a very dangerous thing. Not only may it irritate all philosophers and scholastic theologians, it may also injure the faith and render Holy Scripture false. . . .

If it had been demonstrated with certainty that the sun keeps to the center of the world, that the earth is in the third heaven, that it is not the sun that turns around the earth but the earth that turns around the sun, then one would have to proceed with much circumspection in explicating Scripture. . . . But not until someone has demonstrated this to me will I believe that it exists. It is one thing to prove that by assuming the sun at the center of the world and the earth in the heavens one saves all the appearances and quite another thing to demonstrate that the sun really is at the center, the earth really in the heavens. As to the first, I believe that demonstration can be given; but I have strong doubts as to the second; and in a case of mere doubt you should not diverge from Holy Scripture as the holy fathers have expounded it. . . .

Galileo knew of the letter Bellarmine had written Foscarini: Several writings that were published between the time when he learned of it and his first condemnation contain rebuttals of Cardinal Ballarmine's arguments. Their perusal (Berti was the first to publish excerpts from them) enables us to capture the spirit of Galileo's thought about astronomical hypotheses.

One piece,[26] drawn up towards the end of the year 1615 and addressed to the consultants to the Holy Office, warns them against two errors: the first is to claim that the mobility of the earth is some sort of *tremendous paradox*, an *obvious piece of folly,* not so far demonstrated and for all time indemonstrable; the second is to believe that Copernicus and the

25. This letter was published for the first time by Domenico Berti in *Copernico e le vicende del sistema copernicano in Italia nella seconda metà del secolo XVI e nella prima del secolo XVII* (Rome, 1876), pp. 121–25.

26. Ibid., pp. 132–33.

other astronomers who assumed this mobility "did not believe that it was true in fact and in nature" but admitted it only as a "supposition," to comply with the celestial motions' appearance more easily and to make astronomical calculation more convenient.

In proclaiming that Copernicus believed in the reality of the hypotheses he formulated in the *De revolutionibus,* and in proving (by an analysis of the work) that Copernicus did not admit the earth's mobility and the sun's fixity only *ex suppositione,* as Osiander and Bellarmine would have it, Galileo was upholding the historic truth. But what interests us more than his judgment as a historian is his opinion as a physicist. Now this is easily made out from the piece we are analyzing: Galileo thought, not only that the reality of the earth's motion was demonstrable, but that it had been demonstrated.

This thought stands out still more clearly in another text,[27] from which we learn both that Galileo thought that the Copernican hypotheses are demonstrable, and also how he understood the demonstration to have been carried out:

> Not to believe that the earth's motion is susceptible of demonstration until that demonstration has been exhibited is to act very prudently; nor do we expect anyone to believe such a thing without demonstration. All we would ask is that, for the good of the Holy Church, everything that the followers of this doctrine have produced or everything they would be able to produce be examined with the utmost rigor; let not a single one of their propositions be admitted unless the arguments from which it derives its force far outweigh the reasons of the other party. Let their opinions be rejected if they fail to have more than ninety percent of the reasons on their side. But in return, once it has been proved that the opinion of the philosophers and astronomers of the opposite party is thoroughly false, that it carries absolutely no weight, the opinion of the first party should no longer be sneered at nor should it be given out as so paradoxical that no clear demonstration could conceivably ever be given of it. For the purpose of this debate we can lay down such generous conditions because, clearly, those who hold with the party of error cannot have either reason or experience of any worth on their side, whereas everything must agree and harmonize with the party of truth.
>
> Granted, it is not the same thing to show that on the assumption of the sun's fixity and the earth's mobility the appearances are saved and to demonstrate that such hypotheses are really true in nature. But it should also be granted, and is much more true, that on the commonly accepted system there is no accounting for these appearances, whence this system is indubitably false; just so should it be granted that a system that

27. Ibid., pp. 129–130.

agrees very closely with appearances may be true; and one neither can nor should look for other or greater truth in a theory than this, that it answers to all the particular appearances.

Were one to press this last proposition somewhat, one might easily make it yield the doctrine of Osiander and Bellarmine, that is to say, precisely the one Galileo is attacking. Thus logic constrains the great Pisan geometer to formulate a conclusion directly contrary to the one he had hoped to establish. But earlier in the quoted passage his thought stood out quite clearly.

The pending debate appears to his mind's eye as a sort of duel: Two doctrines, each claiming to be in possession of the truth, announce themselves. The one speaks truly. The other lies. Who will decide between them? Experience! That doctrine with which experience refuses to agree will be recognized as erroneous and, by the same token, the other will be proclaimed to conform to reality. The destruction of one of the two opposing systems guarantees the certainty of the other, just as, in geometry, the absurdity of one proposition entails the truth of its contradictory.

If anyone should doubt that Galileo really held the opinion we are attributing to him, he will be convinced, we believe, by reading the following lines:

The quickest and surest way to show that the position of Copernicus is not contrary to Scripture is, as I see it, to show by a thousand proofs that this proposition is true and that the contrary position cannot be maintained at all. Consequently, since two truths cannot contradict each other, the position recognized as true necessarily agrees with Holy Scripture.[28]

Galileo's notions of the validity of the experimental method and the art of using it are nearly those that Bacon was later to formulate. Galileo conceives of the proof of a hypothesis in imitation of the *reductio ad absurdum* proofs that are used in geometry. Experience, by convicting one system of error, confers certainty on its opposite. Experimental science advances by a series of dilemmas, each resolved by an *experimentum crucis.*

Since this manner of conceiving of the method of experiment was so simple, it was bound to become extremely fashionable; but because it was too simple, it was entirely in error. Grant that the phenomena are no longer saved by Ptolemy's system; the falsity of that system must then

28. Ibid., pp. 105–6.

be acknowledged. But from this it does not by any means follow that the system of Copernicus is true; the latter is, after all, not purely and simply the contradictory of the Ptolemaic system. Grant that the hypotheses of Copernicus manage to save all the known phenomena; that these hypotheses *may* be true is a warranted conclusion, not that they are *assuredly true*. Justification of this last proposition would require that one prove that no other set of hypotheses could possibly be conjured up that would do as well at saving the phenomena. The latter proof has never been given. Indeed, was it not possible, in Galileo's own time, to save all the appearances that could be mustered in favor of the Copernican system by the system of Tycho Brahe?

These logical observations had often been made before Galileo's time. Their justice struck the Greeks the day Hipparchus succeeded in saving the solar motion by either an eccentric or an epicycle. Thomas Aquinas had formulated them with the utmost clarity. Nifo, Osiander, Alessandro Piccolomini, Giuntini—all had repeated them after him. Once again an authoritative voice was to remind the illustrious Pisan of these predecessors.

Cardinal Maffeo Barberini, who was soon to be elevated to the Papacy under the name of Urban VIII, met with Galileo after the condemnation of 1616, to discuss the Copernican doctrine. Cardinal Oregio, present at this meeting, has left us an account of it.[29] At this meeting the future pope, by means of arguments similar to those just rehearsed, laid bare the hidden error of this Galilean argument—since the celestial phenomena all agree with the Copernican hypotheses while they are not saved by the Ptolemaic system, the Copernican hypotheses are certainly true and of necessity in harmony with Holy Writ.

According to Oregio's account, the future Urban VIII advised Galileo:

> to note carefully whether or not there is agreement between the Holy Writ and what he had conceived concerning the earth's motion with an eye to saving the phenomena displayed in the sky and all that philosophers commonly hold as settled by observation and minute scrutiny of what bears on the motions of the heavens and the stars. Granting in effect all that this great scientist had conceived, he [Barberini] asked him whether it was beyond God's power and wisdom to arrange and move the orbs and the stars in a different way while yet saving all the phenomena displayed in the heavens, all that is taught about the stellar motions—their order, position, relative distances, and arrangement.

29. Oregio *Ad suos in universas theologiae partes tractatus philosophicum praeludium completens quatuor tractatus. . . .* (Rome: ex typographia Manelphii, 1637), p. 119. The same account is found on p. 194 of Oregio's *De Deo uno*, written in 1629 (cf. Domenico Berti *Copernico e le vicende*, pp. 138–39).

If you want to maintain that God cannot and knows not how to do this, you must, added the prelate, demonstrate that all these things could not be obtained by a system different from the one you have conceived, that such a system would involve contradiction. For God is capable of all that does not imply contradiction. And since, moreover, God's science is not inferior to his power, if we say that God could have done it, we should also say that he knew it.

If God knew and was able to arrange all things differently from the way you imagined while yet saving all the enumerated effects, then we are not in the least obliged to reduce the divine power and wisdom to this system of yours.

Having heard these words, the great scientist remained silent.

The man who was to become Urban VIII had reminded Galileo of the following truth: no matter how numerous and exact the confirmations by experience, they can never transform a hypothesis into certain truth, for this would require, in addition, demonstration of the proposition that these same experiential facts would flagrantly contradict any other hypotheses that might be conceived.

Were these very logical and prudent admonitions of Bellarmine and Urban VIII sufficient to convince Galileo, to sway him from that exaggerated confidence in the scope of the experimental method, the worth of astronomical hypotheses? One may well doubt it. In the celebrated *Dialogue* of 1632 on the two great world systems Galileo asserts from time to time that he treats the Copernican doctrine as a pure hypothesis without claiming it to be true in nature. But these protestations are given the lie by Salviati's accumulation of proofs in favor of the reality of Copernicus' theory; they are undoubtedly mere pretexts for getting around the interdiction of 1616. At the very moment when the dialogue is about to end, Simplicio, the earnest and dull Peripatetic upon whom devolves the thankless task of defending the system of Ptolemy, concludes with the words:

I confess that your ideas seem much more ingenious to me than many I have heard of; even so, I do not consider them true and conclusive. For I constantly keep a very solid doctrine before my mind's eye, a doctrine I learned from a very learned and eminent person and one before which we must pause. To both of you I want, therefore, to address the following question: Could God in his infinite power and infinite science give to the element of water the oscillating motion that we observe in some other way than by making the containing vessel move? . . . If the answer is yes, I immediately conclude that it would be excessively foolhardy to want to limit and restrict the divine wisdom and power to one particular conjecture.

To which Salviati replies:

An admirable and truly angelic doctrine. One might, in a manner that agrees very closely, answer with another doctrine, one that is divine: Although He permits us to carry on disputes as to the world's constitution, God adds that we are in no condition to discover the work which His hands have made.

Through the mouth of Simplicio and Salviati, Galileo may have hoped to address a delicate piece of flattery to the pope. Perhaps he also wanted to answer the old argument of Cardinal Barberini with a touch of ridicule. This is how Urban VIII took it: To oppose the impenitent realism of Galileo, he gave free reign to the intransigeant realism of the Peripatetics of the Holy Office; the condemnation of 1633 was to confirm the verdict of 1616.

Conclusion

Many philosophers since Giordano Bruno have taken Osiander harshly to task for the preface he placed at the head of Copernicus' book. And Cardinal Bellarmine's and Pope Urban the Eighth's counsels to Galileo have been treated with hardly less severity since the day they were first published.

The physicists of our day, having gauged the worth of the hypotheses employed in astronomy and physics more minutely than did their predecessors, having seen so many illusions dissipated that previously passed for certainties, have been compelled to acknowledge and proclaim that logic sides with Osiander, Bellarmine, and Urban VIII, not with Kepler and Galileo—that the former had understood the exact scope of the experimental method and that, in this respect, Kepler and Galileo were mistaken.

Yet in the history of the sciences Kepler and Galileo are ranked among the great reformers of the experimental method, whereas Osiander, Bellarmine, and Urban VIII are passed over in silence. Is this history's supreme injustice? Could it not be the case that those who ascribed a false scope to the experimental method and who exaggerated its worth worked harder and better at perfecting it than did they whose estimate was from the start more measured and exact?

The Copernicans stubbornly stuck to an illogical realism, although everything drove them to quit that error, and although by ascribing to astronomical hypotheses that "just value" which so many authoritative men had determined for it, they could easily have avoided both the quarrels of philosophers and the censure of theologians. Their conduct is strange indeed and calls for explanation. How can it be explained except by the lure of some great truth—too vaguely apprehended for the Copernicans to be able to enunciate it in its purity, to disengage it from the erroneous contentions that it was hiding under, yet a truth sensed so vividly that neither the precepts of logic nor counsels of prudence could diminish its invisible attraction. What, then, was this truth? This is what we shall now try to articulate.

Throughout antiquity and the Middle Ages, physics displays two divisions so distinct as to be, in a manner of speaking, opposed to each other. On the one hand there is the physics of celestial and imperishable things, on the other the physics of sublunary things subject to generation and corruption.

The beings with which the first of these two kinds of physics deals are regarded as of a nature infinitely higher than that with which the second physics deals; hence the inference that the former is incomparably more difficult than the latter. Proclus teaches that sublunary physics is accessible to man, whereas celestial physics passes his understanding and is reserved for the Divine. Maimonides shares this view of Proclus; celestial physics, according to him, is full of mysteries the knowledge of which God has kept unto Himself; but terrestrial physics, fully worked out, is available in the work of Aristotle.

Yet, contrary to what the men of antiquity and the Middle Ages thought, the celestial physics they had constructed was singularly more advanced than their terrestrial physics.

Ever since the time of Plato and Aristotle, the science of the stars had been organized on the plan which to this day we impose on the study of nature. On the one side there was astronomy—geometers like Eudoxus and Calippus formed mathematical theories by means of which the celestial motions could be described and predicted, while observers estimated to what degree the predictions resulting from calculation conformed to the natural phenomena. On the other side there was physics proper, or to speak in modern terms, celestial cosmology—thinkers like Plato and Aristotle meditated on the nature of the stars and the cause of their movements. What were the relations between these two divisions of celestial physics? What precise line of demarcation was there between them? What affinity united the hypotheses of the one with the conclusions of the other? These questions were debated by the astronomers and physicists of antiquity and the Middle Ages, and they answered them in different ways, for men's minds—then as now—were directed by diverse impulses, impulses very much like those which move modern thinkers.

Much was required before the physics of sublunary things would in its own good time reach a comparable degree of differentiation and organization. In modern times it too will come to be divided into two parts, analogous to those into which celestial physics had been divided since antiquity: the theoretical part combining mathematical systems which by their formulae give knowledge of the exact laws of phenomena; the cosmological part seeking to divine the nature of bodies and their attributes, the nature of the forces to which they are subject or which they exert, the naure of their mutual combinations.

In ancient times, during the Middle Ages, and in the Renaissance, it would have been extremely difficult to make this division. Sublunary physics had but a nodding acquaintance with mathematical theories. Only two branches of that physics—optics (*perspectiva*) and statics (*scientia de ponderibus*)—had the guise of mathematical form, and physicists were hard put to assign them their proper place in the hierarchy of the sciences. Aside from *perspectiva* and *scientia de ponderibus*, analysis of the laws presiding over phenomena remained purely qualitative and rather inexact. Terrestrial physics had not yet freed itself from cosmology.

In dynamics, for instance, the laws of free fall (glimpsed intermittently since the fourteenth century) and the laws of the motion of projectiles (vaguely surmised in the sixteenth century) continued to be involved in metaphysical discussions about local motion, natural and violent motion, coexistence of the mover and the moved. Not until the time of Galileo do we see the theoretical part of physics, whose mathematical form is now being articulated, become disengaged from the cosmological part. Until then the two parts remain intimately united, or rather, inextricably entangled. Their aggregate constituted the physics of local motion.

Meanwhile, however, the ancient distinction between the physics of celestial things and the physics of sublunary things was gradually becoming obliterated. Following Nicholas of Cusa, following Leonardo da Vinci, Copernicus had dared to assimilate the earth to the planets. And Tycho Brahe, by his study of that star which, in 1572, made its appearance and then disappeared, had shown that even the stars are subject to generation and corruption. Galileo, finally, by his discovery of the sun's spots and the moon's mountains, brought the union of the two kinds of physics to completion. Physics was henceforth one science.

When, therefore, Copernicus, Kepler, and Galileo declare with one voice that astronomy should take only propositions whose truth has been established by physics for its hypotheses, this single assertion contains in fact two propositions which are quite distinct. It could be taken to mean that the hypotheses of astronomy are judgments about the nature of heavenly things and their real movements. Or it could signify that the experimental method, by serving as a control on the correctness of astronomical hypotheses, will come to enrich our cosmological knowledge with new truths. The first sense lies, so to speak, at the surface of the assertion. It is immediately manifest. The great astronomers of the sixteenh and seventeenth centuries saw this meaning clearly; they gave formal expression to it, and it was this meaning that solicited their allegiance. Yet so understood, their contention is false and harmful. Osiander, Bellarmine, and Urban VIII rightly viewed it as contrary to logic. It was to engender countless misunderstandings in human science before it was finally rejected.

Beneath this first, illogical, but manifest and seductive meaning there lay another: in demanding that the hypotheses of astronomy accord with the teachings of physics, the Renaissance astronomers were in effect requiring that the theory of the celestial motions rest upon bases that could support the theory of the motions we observe here below as well. The courses of the stars, the ebb and flow of the sea, the motion of projectiles, the fall of heavy bodies—*all* were to be saved by *one and the same* set of postulates, postulates formulated in the language of mathematics.

This meaning remained deeply hidden. Not Copernicus, or Kepler, or Galileo saw it clearly. Beneath the clear but false and dangerous meaning which the Renaissance astronomers had seized upon, the other, though disguised, retained its fertility. And while the false and illogical sense which they ascribed to their principle engendered disputes and quarrels, the true but hidden meaning of this same principle gave birth to the scientific efforts of these inventors. While straining to prop up the strict truth of the former, they were, unknowingly, establishing the correctness of the latter. Kepler, when he tried again and again to give an account of the motions of the stars in terms of the properties of water currents or of magnets, Galileo, when he attempted to make the path of projectiles accord with the motion of the earth or when he tried to derive an explanaion of the tides from the earth's motion—both believed that they were thus proving that the Copernican hypotheses have their foundation in the nature of things. But the truth which, little by little, they were introducing into science was that one form of dynamics, by means of a single set of mathematical formulae, must represent the movements of the stars, the oscillations of the sea, the fall of heavy bodies. They thought they were "renovating" Aristotle; in fact they were preparing for Newton.

Despite Kepler and Galileo, we believe today, with Osiander and Bellarmine, that the hypotheses of physics are mere mathematical contrivances devised for the purpose of saving the phenomena. But thanks to Kepler and Galileo, we now require that they save *all the phenomena* of the inanimate universe *together.*

Index